Springer Proceedings in Mathematics & Statistics

Volume 500

This book series features volumes composed of selected contributions from workshops and conferences in all areas of current research in mathematics and statistics, including data science, operations research and optimization. In addition to an overall evaluation of the interest, scientific quality, and timeliness of each proposal at the hands of the publisher, individual contributions are all refereed to the high quality standards of leading journals in the field. Thus, this series provides the research community with well-edited, authoritative reports on developments in the most exciting areas of mathematical and statistical research today.

Arthur Gibadullin • Sergey Gordin
Editors

Computing Technologies and Applied Mathematics

CTAM 2024, Komsomolsk-na-Amure, Russia, October 07-11, 2024

Springer

Editors
Arthur Gibadullin
National Research University "Moscow Power Engineering Institute"
Moscow, Russia

Sergey Gordin
Komsomolsk-na-Amure State University
Komsomolsk-on-Amur, Russia

ISSN 2194-1009 ISSN 2194-1017 (electronic)
Springer Proceedings in Mathematics & Statistics
ISBN 978-3-031-90545-2 ISBN 978-3-031-90546-9 (eBook)
https://doi.org/10.1007/978-3-031-90546-9

Mathematics Subject Classification: 90C05; 49M05; 00A69; 00A71; 90B18

This Springer imprint is published by the registered company Springer Nature Switzerland AG
The registered company address is: Gewerbestrasse 11, 6330 Cham, Switzerland

Preface

The conference was held to summarize international experience in the field of mathematics, information, and computing technologies. The III International Seminar on Computing Technologies and Applied Mathematics was held on October 07–11, 2024, in person at Komsomolsk-na-Amure State University (Komsomolsk-on-Amur, Russia). The conference addressed issues of mathematics related to mathematical and optimization modeling, numerical algorithms, and high-performance computing and applications. A distinctive feature of the conference is that authors from Germany, China, South Korea, Uzbekistan, and Russia spoke at it. Researchers from different countries presented scientific and practical proposals on mathematical modeling and computing technologies.

The editors of the conference proceedings are Arthur Gibadullin from the National Research University "MPEI," Moscow, Russia, and Sergey Gordin from Komsomolsk-na-Amure State University, Komsomolsk-on-Amur, Russia. Thus, the conference allowed developing new scientific recommendations on the use of mathematics, information, computer, digital and intelligent technologies, and networks in various fields of activity, which can be useful for state and regional authorities, international and supranational organizations, and scientific and professional community.

Each presented paper was reviewed by at least three members of the Program Committee or an independent reviewer. As a result of the work of all reviewers, 14 articles out of 37 received materials were accepted for publication. The reviews were based on the assessment of the topic of the presented materials, relevance of the research, scientific significance and novelty, quality of materials, and originality of the work. Reviewers, members of the Program Committee, and members of the Organizing Committee did not enter into discussions with the authors of the articles.

The Organizing Committee of the conference expresses its gratitude to the Springer staff that supported the publication of this collection. In addition, the Organizing Committee thanks the conference participants, reviewers, and everyone who helped organize this conference and form this volume for publication in the Springer Proceedings in Mathematics & Statistics series.

Moscow, Russia
Komsomolsk-on-Amur, Russia

Arthur Gibadullin
Sergey Gordin

Organization

Program Committee Chairs	
Alekseev Gennadii	Far Eastern Federal University, Vladivostok, Russia
Chebotarev Alexander	Far Eastern Federal University, Vladivostok, Russia
Guzev Michael	Institute of Applied Mathematics FEB RAS, Vladivostok, Russia
Program Committee	
Samardak Alexander	Far Eastern Federal University, Vladivostok, Russia
Kovtanyuk Andrey	Technische Universität München, Munich, Germany
Nurminski Evgeni	Far Eastern Federal University, Vladivostok, Russia
Dmitriev Eduard	Komsomolsk-na-Amure State University, Komsomolsk-on-Amur, Russia
Komarov Oleg	Institute of Metallurgy and Mechanical Engineering FEB RAS, Komsomolsk-on-Amur, Russia
Liu Chi	Beijing Institute of Technology, Beijing, China
Kulchin Yuri	Far Eastern Branch, Russian Academy of Sciences, Vladivostok, Russia
Xu Runzhang	Harbin Engineering University, Harbin, China
Marfin Yuriy	Pacific National University, Khabarovsk, Russia
Sushchenko Andrei	Institute of Applied Mathematics FEB RAS, Vladivostok, Russia
Amiraliyev Gabil	Erzincan Binali Yıldırım Üniversitesi, Erzincan, Turkey
Chigarev Anatoly	Belarusian State University, Minsk, Belarus
Devi Geetha	National University of Science and Technology, Muscat, Oman
Chandrasekhar Garlapati	Puducherry Technological University, Puducherry, India
Uzakov Gulom	Karshi State Technical University, Karshi, Uzbekistan

(continued)

Dimitrov Lubomir	Technical University of Sofia, Sofia, Bulgaria
Britvina Valentina	Moscow Polytechnic University, Moscow, Russia
Sadriddinov Manuchehr	Small Academy of Sciences under the Presidium of the National Academy of Sciences of Tajikistan, Dushanbe, Tajikistan
Gibadullin Arthur	National Research University "Moscow Power Engineering Institute," Moscow, Russia, and Financial University under the Government of the Russian Federation, Moscow, Russia
Voinash Sergey	Kazan Federal University, Kazan, Russia
Morkovkin Dmitry	Financial University under the Government of the Russian Federation, Moscow, Russia
Gordin Sergey	Komsomolsk-na-Amure State University, Komsomolsk-on-Amur, Russia
Pulyaeva Valentina	Financial University under the Government of the Russian Federation, Moscow, Russia
Mehdiyeva Almaz	Azerbaijan State University of Oil and Industry, Baku, Azerbaijan
Khamraeva Sayyora	Karshi State Technical University, Karshi, Uzbekistan
Organizer	
Komsomolsk-na-Amure State University, Komsomolsk-on-Amur, Russia	

Contents

Information-Mathematical Support of Digital Twin of Stand-Alone Hybrid Energy System

V. D. Berdonosov, O. N. Feygenson, and E. O. Shelopugina

Abstract The paper presents mathematical models of solar energy fluxes considering cloud cover, data model from the experimental unit, and energy balance model. Also, the method of data rectification due to experimental measurements is described. It proposes the configuration of a digital twin of an isolated hybrid energy system that includes three channels: simulation (theoretical), hardware (physical) and secondary. Preliminary results of isolated hybrid energy system functioning are analysed. The developed system allows qualitative and quantitative analyses of energy flows, which contributes to more efficient management of energy resources and improved reliability of energy supply, in particular by detecting critical time periods with energy shortages.

Keywords Mathematical models · Digital twin · Isolated hybrid energy system · Energy flows · Energy supply · Energy shortage

1 Introduction

Current level of technology development leads to increased needs for power supply to many autonomous objects, whose geographical location makes it impossible to connect them to centralised power supply systems. Such conditions cause the relevance and practical value of studying ways to cover the entire energy needs of local areas isolated from centralised energy supply systems by using available natural renewable sources.

Usual hybrid energy systems (HES) utilising renewable energy sources (RES) are considered to replace conventional energy systems in industry, homes, public

V. D. Berdonosov · E. O. Shelopugina
Komsomolsk-na-Amure State University, Komsomolsk-na-Amure, Russia

O. N. Feygenson (✉)
Samsung Electronics, Suwon, South Korea
e-mail: on.feygenson@samsung.com

A. Gibadullin, S. Gordin (eds.), *Computing Technologies and Applied Mathematics*, Springer Proceedings in Mathematics & Statistics 500,
https://doi.org/10.1007/978-3-031-90546-9_1

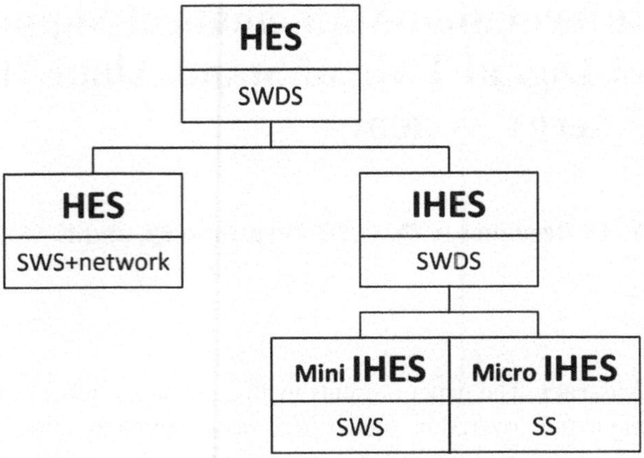

Fig. 1 HES hierarchies: *S* solar panel, *W* wind generator, *D* diesel generator, *S* storage

Table 1 Amount of data related to the sampling

Sampling level	Data size per variable
Month	12
Day	365
Hour	8760
10 min	52,560

facilities and so on [1]. In general, these include: solar panels, wind generators, diesel generators, storage, and distribution and control equipment [2]. Such HES can be integrated into the common power network, thus eliminating the need for additional generators, e.g. diesel generators [3], or autonomously operated, i.e. isolated HES (IHES) [4–6].

In turn, IHES can include both full-fledged systems and mini and micro-HES. Mini-HES means systems with a minimal set of components: a solar panel, a wind generator and storage, e.g. lights in public areas. Micro-HES include systems with a super-minimal set of elements: solar panel and storage, e.g. power for traffic cameras, power for traffic lights, illuminated crosswalks [7] (Fig. 1).

This paper only considers micro-IHES consisting of solar panels, storages and information and control infrastructure (controllers, indicators and so on).

For usual HES there are a large number of studies on generalised economic optimisation [8], and these studies are based on monthly [9], daily and hourly data, at least for a year [7]. In general, 10-min data may also be of interest, but no such studies have yet been found in the literature. The frequency of data representation will be referred to as sampling or sampling level. Table 1 shows the data size related to the sampling. Note that the optimisation requires data on at least solar radiation, wind speed, temperature, consumption levels etc. That is, there must be at least four times as much data.

Such a large amount of data available for optimisation requires the most efficient algorithms for large data sets to be used to achieve acceptable accuracy [10]. Such algorithms, for example, include: hybrid spotted hyena optimisation algorithm

Table 2 Amount of data related to the sampling

Sampling level	Data size per variable for a year	Data size per variable for 10 days
Month	12	–
Day	365	10
Hour	8760	240
10 min	52,560	1440

(HSHOA), ant colony optimisation (ACO), cuckoo search algorithm (CSA), moth flame optimisation (MFO), genetic algorithm (GA), spotted hyena optimisation (SHO), grey wolf optimisation (GWO), grasshopper optimisation algorithm (GOA) and particle swarm optimization (PSO) [10].

Thus, there is a contradiction in optimisation: with increasing accuracy (reliability) by higher sampling rates, optimisation procedures become unacceptably complex due to the large size of the dataset. The contradiction is caused by stochastic nature of the solar energy flux. Stochasticity is mainly related to significant variations in cloud cover and, to a lesser extent, to variations in albedo.

This contradiction can be resolved by TRIZ [11] tools of contradiction elimination, for example, by the method of Local Quality. Following this method, the optimisation shall not be carried out for the entire annual period, but only for the time period with the worst ratio of incoming solar energy flow to the energy flow delivered to a consumer. Let us call this interval as a critical time period. Under this approach, the time period is reduced from a year to a few days. Assuming that the duration of the critical time period does not exceed 10 days, we obtain the following dataset (see Table 2).

It should be noted that the use of a critical time period instead of a full, annual time period is unconditionally valid only for systems with deterministic consumption. However, for most micro-HES, consumption is exactly deterministic.

Optimizing for both the full annual period and the critical time period, there is a data acquisition problem. These data can be obtained either theoretically or experimentally using some experimental (pilot) unit. The experimental unit gives valid data, but requires a large amount of additional resources. Theoretical approach allows to significantly reduce the amount of resources required, but data reliability decreases. That is, there is a second contradiction: with increasing reliability, the amount of resources consumed increases unacceptably. This contradiction can be eliminated by the method of Copying. To apply this method, an approach that implements a digital twin shall be used.

Let us discuss the current state-of-art of researches related to digital twins in the field of energy. In [12], a six-layer digital twin design is proposed for large energy facilities, comprising ontology model, digital circuitry, electronic documentation, facility information model, operational information from instrumentation and sensors, mathematical and simulation models. For the given task, the facility is much simpler, so we can limit ourselves to the layer "mathematical and simulation models" with the inclusion of elements from the layer "operational information from measuring devices and sensors". Furthermore, it discusses the creation of digital and smart digital twins for modelling and control in power systems, analysing integrated smart power systems, trends in power industry digitalisation and the application of ontology approach and fractal method in ontology design, with examples of digital twin implementation for solar power plant and smart digital twin for fuel and energy complex [13].

Some digital twin developers use the data from smart energy meters and additional building data including roof shape, material, wall thickness and so on [14]. There is also an information on digital twins using deep learning artificial neural networks for intelligent control [15]. The literature [16] notes that predictive digital twins can be implemented using different types of modelling, including both methods based on the physics of processes and using machine learning algorithms. The need for a comprehensive and robust digital twin design for modern power systems is emphasised, noting the need to apply physical and mathematical models, statistical data analysis and real data to improve reliability and availability [17]. There are studies on key modelling techniques such as, agent-based modelling (ABM), system dynamics (SD), discrete event simulation (DES) and integrated energy models (IEM) [8]. It should be noted that there are systematic and critical reviews of digital twin technology in the power industry, emphasising its potential to improve power management maintenance and integration of renewable energy resources [18].

Thus, considering the previous studies, it can be indicated that the digital twin of a solar energy system is a virtual model that exactly reflects the physical structure and operation of a real solar facility. It considers various parameters such as geographical location, weather conditions, solar panel characteristics and electrical components. Using the digital twin, various scenario analyses can be carried out to determine the optimal operating conditions for solar systems. For example, the effects of different solar panel angles, solar radiation direction and intensity on system performance can be evaluated. This allows management strategies to be developed to maximise energy production.

One of the main digital twin advantages is the ability to perform virtual testing and analysis without the need for physical resources. This significantly reduces the time and cost of a study, and allows different scenarios and strategies to be explored without risking the real system. The digital twin also allows prediction of solar system performance based on current conditions and past experience.

So, summarizing the above, we can formulate the research task: to develop an algorithm to identify the critical time period for micro IHES based on cloud level data using digital twin.

To solve the given task, a digital twin has to be developed first of all. The digital twin (DT) of input energy flows of an isolated hybrid energy system (IHES) contains three channels: simulation (theoretical), hardware (physical) and secondary (Fig. 2).

The theoretical channel simulates a solar energy flux in ClearSky mode, assuming no cloud cover but taking albedo into account.

The secondary channel receives Cloudiness data, which is then used to adjust solar fluxes in ClearSky mode. This provides the solar energy flux (SolarFlux_Cl) that most closely matches the real one.

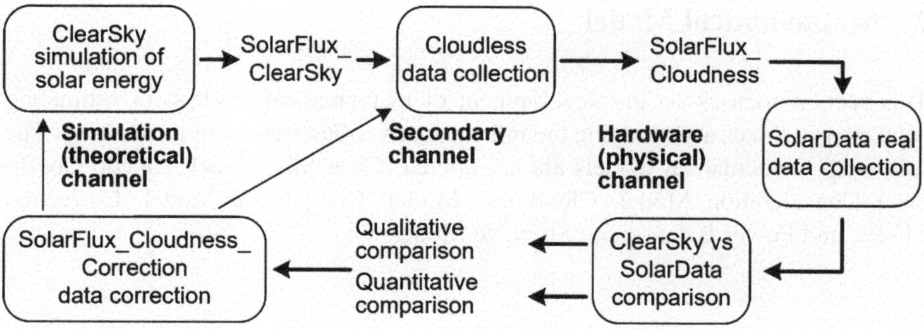

Fig. 2 IHES DT channels

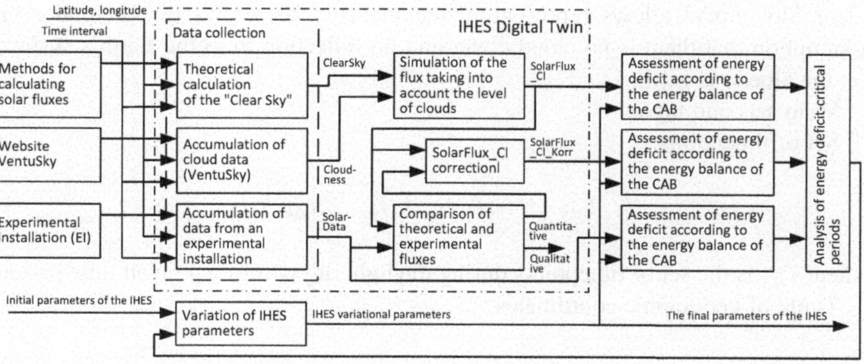

Fig. 3 Architecture diagram of IHES DT

The physical channel captures solar flux data from a real facility (SolarData). ClearSky and SolarData are compared to identify qualitative and quantitative differences as the data becomes available in each channel. Qualitative comparison allows assessing the consistency of cloud data obtained from an external source (e.g. VentuSky website) with that of the real facility. Quantitative differences are averaged to obtain a correction coefficient. This coefficient is then used to correct SolarFlux_Cl data forming SolarFlux_Cl_Korr.

Thus, the digital twin (DT) generates three data flows: SolarFlux_Cl, SolarFlux_Cl_Korr and SolarData. Each of these flows is fed to the power shortage estimation unit of IHES battery, which consider IHES parameters such as solar panel area and efficiency, battery capacity and power consumption level. Figure 3 shows the architecture diagram of IHES DT and its external interface parts.

Then, in order to develop a general mathematical model of the given task, it is necessary to create a mathematical model for each part of IHES DT configuration.

2 Mathematical Model

This section focuses on the development of mathematical models for estimating solar energy fluxes and studying the influence of various factors on its intensity. The following mathematical models are considered: Clear Sky Model, Albedo Coefficient Consideration Model, Cloudiness Model, Comparison Model, Correction Model, and Power Balance and Shortage Model.

2.1 Clear Sky Model

Clear Sky model allows representing the intensity of solar radiation under the assumption that there is no cloud cover and no reflection from the Earth's surface, i.e. the albedo is zero.

Main sets and tuples:

Set of timepoints:

$$T_{sky} = \{t_1, t_2, t_3, \ldots, t_{sky}, \ldots t_{nsky}\} \tag{1}$$

where T_{sky} is the set of timepoints during daylight days, during a given time period.

Tuple of geographic coordinates:

$$L = (latitude, longitude) \tag{2}$$

where *latitude* is latitude; *longitude* is longitude.

Set of observer parameters:

$$O(date) = \{L, timezone, date\} \tag{3}$$

where L is the tuple of geographic coordinates; *timezone* is the time zone; *date* is the date of observation.

Tuple of solar time parameters:

$$S(t_{sky}) = \{zenith\_angle(t_{sky}), solar\_flux(t_{sky})\} \tag{4}$$

where $zenith\_angle(t_{sky})$ is the zenith angle at time t_{sky}; $solar\_flux(t_{sky})$ is the solar flux at time t_{sky}.

Main functions:

Function to get the time of sunrise and sunset:

$$Sunrise\_Sunset(O(date)) \rightarrow (sun\_rise(date), sun\_set(date)) \tag{5}$$

where *Sunrise _ Sunset* is a function that takes observer parameters and returns the time of sunrise and sunset.

Function to create a time range:

$$Time\_Range(sun\_rise, sun\_set) \rightarrow T_{sky} \tag{6}$$

where *Time _ Range* is a function that creates a time range between sunrise and sunset.

Function to calculate a zenith angle:

$$Zenith\_Angle\left(t_{sky}, L\right) \rightarrow \theta\left(t_{sky}\right) \tag{7}$$

where *Zenith _ Angle* is a function taking timepoint t_{sky} and coordinates L, and returning the zenith angle. $\theta(t_{sky})$.

Function for calculating the distance between the Earth and the Sun:

$$Earth\_Sun\_Distance\left(t_{sky}\right) \rightarrow r\left(t_{sky}\right) \tag{8}$$

where *Earth _ Sun _ Distance* is a function returning the distance between the Earth and the Sun at time t_{sky}.

Function for calculating the solar energy flux:

$$Solar\_Flux\left(\theta\left(t_{sky}\right), r\left(t_{sky}\right)\right) \rightarrow E_{strim}^{clear}\left(t_{sky}\right) \tag{9}$$

where *Solar _ Flux* is a function taking zenith angle $\theta(t_{sky})$ and distance $r(t_{sky})$, and returning solar energy flux $E_{strim}^{clear}\left(t_{sky}\right)$.

The model describes a sequence of steps and dependencies to calculate the solar energy flux at a given geographic coordinate.

2.2 Albedo Coefficient Consideration Model

This model considers how the reflection from the Earth's surface affects the final solar radiation intensity at given timepoints. The subsection uses sets and tuples defined above.

Main functions:

Function for calculating albedo coefficient:

$$Albdo\_Coefficient(date) \rightarrow Albedo(\mathrm{date}) \tag{10}$$

where *Albedo _ Coefficient* is a function taking the day number of a year and returning albedo coefficient.

Function to calculate the solar energy flux based on albedo:

$$Solar\_Flux\_Albedo(\theta(t_{sky}), r(t_{sky}), Albedo(date)) \rightarrow E_{strim}^{albedo}(t_{sky}) \quad (11)$$

where $Solar\_Flux\_Albedo$ is a function taking zenith angle $\Theta(t_{sky})$, distance $r(t_{sky})$ and albedo coefficient, and returning the solar energy flux $E_{strim}^{albedo}(t_{sky})$.

The model describes a sequence of steps and dependencies to calculate the solar energy flux considering albedo coefficient at a given geographic coordinate.

2.3 Cloudiness Model

This model accounts for the effect of cloud cover on the final solar radiation intensity at given timepoints.

Main sets and tuples:

Tuple of website request parameters:

$$Q(date) = (L, start\_time, end\_time, data\_type) \quad (12)$$

where L is a tuple of geographic coordinates; $start\_time$ is a start time and end_time is an end time specifying a date range; $data\_type$ is requested data kind, in this case it is cloud cover.

Set of solar radiation and cloud cover parameters:

$$M(t_{sky}) = \{solar\_rad(t_{sky}), cloudiness(t_{sky})\} \quad (13)$$

where $solar\_rad$ is solar radiation level (W/m^2); $cloudiness$ is cloudiness level in per cent (%).

Tuple of system state parameters:

$$E(t_{sky}) = (solar\_rad(t_{sky}), cloudiness(t_{sky}), calculated\_solar\_rad(t_{sky})) \quad (14)$$

where $E(t_{sky})$ are system parameters at time t_{sky}.

Main functions:

Function to generate a database query where cloudiness data are stored:

$$Generate\_Request(Q(date)) \rightarrow URL \quad (15)$$

where $Generate\_Request$ is a function that generates URL to request cloudiness data.

Function for sending a request and getting data:

$$Fetch\_Data(URL) \rightarrow C(t_{sky}) \quad (16)$$

where *Fetch _ Data* is a function that sends a request to the database and returns cloudiness data.

Function to collect data:

$$Collect\_Data\left(T_{sky}\right) \rightarrow \left\{\left(solar\_rad\left(t_{sky}\right), cloudiness\left(t_{sky}\right)\right) \mid t_{sky} \in T_{sky}\right\} \quad (17)$$

where *Collect _ Data* is a function that collects solar radiation and cloudiness data at each timepoint t_{sky}.

Function to prevent division by zero and replace NaN:

$$Sanitize\_Data\left(M\left(t_{sky}\right)\right) \rightarrow M_{sanitized}\left(t_{sky}\right) \quad (18)$$

where *Sanitize _ Data* is a function that cleans data from NaN and prevents division by zero.

Function for calculating solar energy with regard to cloudiness:

$$Calculate\_Solar\_Energy\_Cloudiness\left(solar\_rad\left(t_{sky}\right), cloudiness\left(t_{sky}\right)\right) \rightarrow E_{cloudiness}\left(t_{sky}\right) \quad (19)$$

where *Calculate _ Solar _ Energy _ Cloudiness* is a function that calculates solar energy with regard to cloudiness.

The model describes collection and analysis of solar radiation and cloudiness data to calculate cloud-adapted solar energy. It includes functions for data collection, data cleaning and calculation of cloud-adapted solar radiation.

2.4 Comparison Model

A mathematical model for comparing the results assumes a comparison of solar energy data simulated with respect to cloudiness with experimental data obtained from a pilot unit.

Main sets and tuples:

Tuple of coordinates and system parameters from an experimental unit:

$$C\_ExpU = (L, timezone) \quad (20)$$

where L is a tuple of geographic coordinates; *timezone* is a time zone.

Set of monitoring parameters:

$$M\_comp\left(t_{sky}\right) = \{solar\_rad, temp, humidity, energy\_production, energy\_consumption, equipment\_status\} \quad (21)$$

where *solar _ rad* is solar radiation level (W/m^2); *temp* is air temperature (°C); *humidity* is air humidity (%); *energy _ production* is power generation (kWh); *energy _ consumption* is power consumption (kWh); *equipment _ statuse* is equipment condition (nominal values).

Tuple of system state parameters:

$$E(t_{sky}) = \Big(solar\_rad(t_{sky}), temperature(t_{sky}), humidity(t_{sky}), \\ energy\_production(t_{sky}), energy\_cosumtion(t_{sky}), \\ equipment\_status(t_{sky})\Big) \quad (22)$$

where $E(t_{sky})$ are system parameters at time t_{sky}.

Set of cloud-adapted solar energy modelling data:

$$M\_mod = \{(time_1, solar\_energy_1), (time_2, solar\_energy_2), \ldots, (time_n, solar\_energy_n)\} \quad (23)$$

where *M_mod* is the set of cloud-adapted simulated solar energy data, where each element contains a timepoint and modelled solar energy value.

Set of experimental data on solar energy:

$$\mathrm{E\_exp\_U} = \{(\mathrm{time}_1, \mathrm{exp\_U\_energy}_1), (\mathrm{time}_2, \mathrm{exp\_U\_energy}_2), \ldots, (time_n, exp\_U\_energy_n)\} \quad (24)$$

where *E_exp_U is* the set of experimental solar energy data, where each element contains a timepoint and experimental solar energy value.

Main functions:

Function to collect data:

$$Collect\_Data(T_{sky}) \to \{E(t_{sky}) \mid t_{sky} \in T_{sky}\} \quad (25)$$

where *Collect _ Data* is a function that collects data at each timepoint t_{sky}.

Function for calculating the efficiency of a solar panel:

$$Efficiency\_SP(solar\_rad, energy\_production) \to \eta_{SP} \quad (26)$$

where *Efficiency _ SP* is a function that calculates the efficiency of a solar panel based on the level of solar irradiation and power generation.

Function to calculate the effect of temperature and humidity on system efficiency:

$$Climate\_Impact(temperature, humidity, \eta SP) \to \eta_{adjusted} \quad (27)$$

where *Climate* _ *Impact* is a function that corrects the efficiency of a solar panel based on temperature and humidity.

Function for analyzing power balance:

$$Energy\_Balance(energy\_production, energy\_consumption) \rightarrow balance \quad (28)$$

where *Energy* _ *Balance* is a function that calculates balance between energy production and consumption.

Function for monitoring equipment condition:

$$Equipment\_Status\_Analysis(equipment\_status) \rightarrow status\_report \quad (29)$$

where *Equipment_Status_Analysis* is a function that analyses equipment condition and generates a report.

Function for comparing simulated data:

$$Compare\_Data(M, E) \rightarrow R \quad (30)$$

where *Compare* _ *Data* is a function comparing simulated and experimental solar energy data; R *is* the comparison result, as, for example, mean square error or determination coefficient.

Model of experimental data describes the collection and analysis of experimental data of an isolated hybrid energy system (IHES). It comprises functions for evaluating solar panel efficiency, considering climatic conditions, analysing power balance and monitoring equipment condition. This model can be used to assess the accuracy of the modelling and to compare the results with actual measurements.

2.5 Correction Model

Mathematical model for solar radiation data processing allows filtering data by periods, merging them, calculating correction coefficient and correcting solar flux values.

Main sets and tuples:

Set of solar flux data:

$$F = \{(datetime_1, solar\_flux_1), (datetime_2, solar\_flux_2), \ldots, (datetime_n, solar\_flux_n)\} \quad (31)$$

where F is a set of solar flux data, where each element contains a timepoint and solar flux value.

Set of solar radiation data:

$$R = \{(datetime_1, solar\_rad_1), (datetime_2, solar\_rad_2), \ldots, (datetime_m, solar\_rad_m)\} \quad (32)$$

where R is a set of solar radiation data, where each element contains a timepoint and solar radiation value.

Main functions:

Function for filtering data by period:

$$filter\_data(start, end, D) \rightarrow D' \quad (33)$$

where *filter _ data* is the function that filters data on given time frame; *start* is an initial timestamp; *end* is a final timestamp; D is a data set; D' is a filtered data set satisfying conditions $start \le datetime \le end$.

Function for merging data by timestamps:

$$merge\_data(F', R') \rightarrow M \quad (34)$$

where *merge _ data* is a function that merges data from sets F' and R' by timestamp; $F' \subseteq F$ is a filtered set of solar flux data; $R' \subseteq R$ is a filtered set of solar radiation data; $M \subseteq \mathbb{T} \times \mathbb{R} \times \mathbb{R}$ is merged data set, where each element contains a timestamp, solar flux value, and solar radiation value.

Function for calculating correction coefficient:

$$calculate\_correction\_coefficient(M) \rightarrow k \quad (35)$$

where *calculate _ correction _ coefficient* is a function that calculates the correction coefficient based on data from set M; M is the set of merged data; $k \in \mathbb{R}$ is the correction coefficient.

Function for correcting solar flux values:

$$correct\_solar\_flux(F', k) \rightarrow F'' \quad (36)$$

where *correct _ solar _ flux* is a function that rectifies solar flux values based on a correction factor; $F' \subseteq F$ is filtered solar flux data set; $k \in \mathbb{R}$ is the correction factor; $F'' \subseteq F$ is rectified solar flux data set.

This model is the basis for data processing algorithm and designing appropriate software solutions.

2.6 Power Balance and Shortage Model

This model defines expressions to calculate power balance and shortage in the system of solar power generation and battery storage.

Main sets and tuples:
Set of solar flux and cloudiness data:

$$C = \Big\{ (datetime_1, solar\_flux_1, cloudiness_1), \\ (datetime_2, solar\_flux_2, cloudiness_2), \ldots, \\ (datetime_n, solar\_flux_n, cloudiness_n) \Big\} \quad (37)$$

where C is a set of solar flux and cloud cover data, where each element contains a timepoint, solar flux value and cloudiness value.

Main functions:

Function for filtering data by period:

$$filter\_data(start, end, C) \rightarrow C' \quad (38)$$

where $filter\_data$ is the function that filters data on given time frame; $start$ is an initial timestamp; end is a final timestamp; D is a data set; D' is a filtered data set satisfying conditions $start \leq datetime \leq end$.

Function for time rounding to the nearest hour:

$$round\_time(C') \rightarrow C'' \quad (39)$$

where $round\_time$ is a function that rounds time to the nearest hour; $C' \subseteq C$ is a filtered data set; $C'' \subseteq C$ is the data set with rounded time.

Function for zeroing solar flux values below the threshold:

$$nullify\_low\_flux(C'', minSolar) \rightarrow C''' \quad (40)$$

where $nullify\_low\_flux$ is a function that zeroes solar flux values below a given threshold; $C'' \subseteq C$ is a data set with rounded time; $minSolar \in \mathbb{R}$ is a solar flux threshold value; $C''' \subseteq C$ is a data set with zeroed solar flux values.

Function to create a full-time range:

$$create\_full\_range(C''') \rightarrow T_{dif} \quad (41)$$

where $create\_full\_range$ is a function that creates a full time range with 60 min intervals; $C''' \subseteq C$ is a data set; T_{dif} is a set of time periods.

Function for merging data by timestamp:

$$merge\_data(T_{dif}, C''') \rightarrow C_{full} \quad (42)$$

where $merge\_data$ is a function that merges data by a timestamp;

T_{dif} is the set of time periods; $C''' \subseteq C$ is a data set; $C_{full} \subseteq C$ is a merged data set.

Function for linear interpolation of missing values:

$$interpolate\_data\left(C_{full}\right) \rightarrow C_{interp} \tag{43}$$

where *interpolate _ data* is the linear interpolation function of missing values; $C_{full} \subseteq C$ is merged data set; $C_{interp} \subseteq C$ is a data set with interpolated values.

Function for calculating the instantaneous power of a solar panel:

$$calculate\_instant\_\ power\left(C_{interp}, photoEfficiency, photoCellSquare\right) \rightarrow C \tag{44}$$

where *calculate_instant_power* is the function to calculate the instantaneous power of a solar panel; $C_{interp} \subseteq C$ is a data set with interpolated values; $photoEfficiency \in \mathbb{R}$ is solar panel efficiency; $photoCellSquare \in \mathbb{R}$ is solar panel area; $C_{power} \subseteq C$ is a data set with calculated instantaneous power.

Function for calculating stored solar energy:

$$calculate\_solar\_energy\left(C_{power}\right) \rightarrow C_{energy} \tag{45}$$

where *calculate_solar_energy* is a function for calculating stored solar energy; $C_{power} \subseteq C$is a dataset with calculated instantaneous power; $C_{energy} \subseteq C$ is a dataset with calculated stored solar energy.

Function for calculating battery power flows:

$$calculate\_battery\_\ power\left(C_{energy}, load\_\ power, charge\_voltage, maxChargeCurrent\right) \rightarrow C_{battery} \tag{46}$$

where *calculate _ battery _ power* is a function for calculating battery power flows; $C_{energy} \subseteq C$ is a data set of stored solar energy; $load\ \_\ power \in \mathbb{R}$ is power consumption; $charge\ \_\ voltage \in \mathbb{R}$ is charging voltage; $maxChargeCurrent \in \mathbb{R}$ is maximum charging current; $C_{battery} \subseteq C$ is a data set of calculated battery power flows.

Function for calculating accumulated energy in a battery:

$$calculate\_battery\_energy\left(C_{battery}, batteryCapacity, system\_voltage, start\_energy\right) \rightarrow C_{balance} \tag{47}$$

where *calculate _ battery _ energy* is the function for calculating accumulated energy in a battery; $C_{battery} \subseteq C$ is a data set of calculated battery power flows;

$batteryCapacity \in \mathbb{R}$ is battery capacity; $system\ \_\ voltage \in \mathbb{R}$ is system voltage; $start\ \_\ energy \in \mathbb{R}$ is initial power value; $C_{balance} \subseteq C$ is a data set of calculated accumulated energy in a battery.

Function for calculating power shortage:

$$calculate\_energy\_shortage(C_{balance}, DoD) \rightarrow C_{shortage} \tag{48}$$

where $calculate\_energy\_shortage$ is a function for calculating power shortage; $C_{balance} \subseteq C$ is a data set of calculated accumulated energy in a battery; $DoD \in \mathbb{R}$ is discharge rate; $C_{shortage} \subseteq C$ is a data set of calculated power shortage.

Mathematical model for calculating power balance and shortage in a system of solar generation and battery storage allows filtering data, rounding time, zeroing minor values, interpolating missing values, calculating instantaneous and accumulated energy as well as power shortage.

3 Simulation Results

The architecture diagram of micro IHES monitoring system, shown in Fig. 4, includes several key components. Sensors are designed to collect a variety of information on the state of IHES equipment, including temperature, humidity, solar radiation level, and operating status of the equipment. The controller, in turn, collects and processes data from the battery and solar panel, recording it in its software running on the Windows operating system. Hereinafter this system is referred to as the experimental unit.

Experimental data obtained in the study is a multivariate set of information containing observations and measurements of various micro IHES performance parameters.

Solar radiation. Measurements of solar radiation levels under different daylight conditions, measured in watt-hours per square metre (W/m^2), allow assessing the efficiency of solar panels and determining the dependence of power generation on the intensity of sunlight.

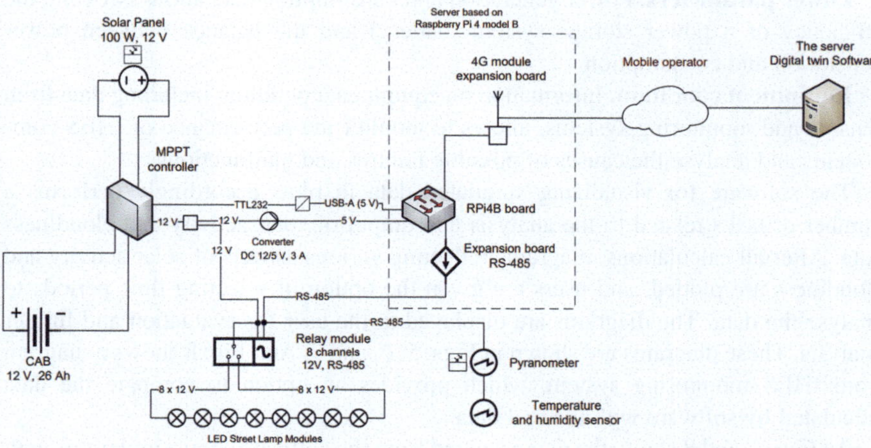

Fig. 4 Architecture diagram of micro IHES monitoring system (experimental unit)

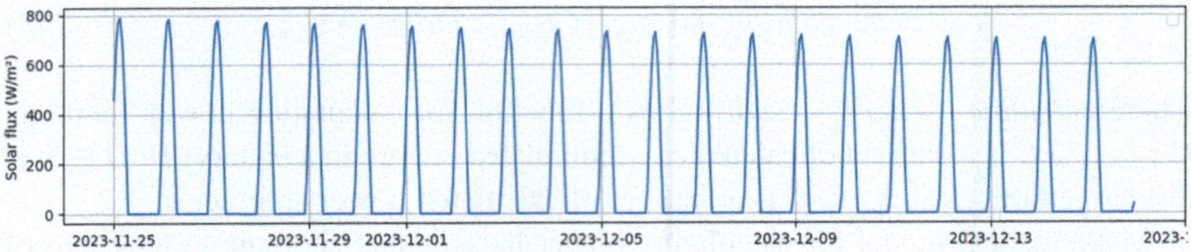

Fig. 5 Solar flux intensity considering albedo

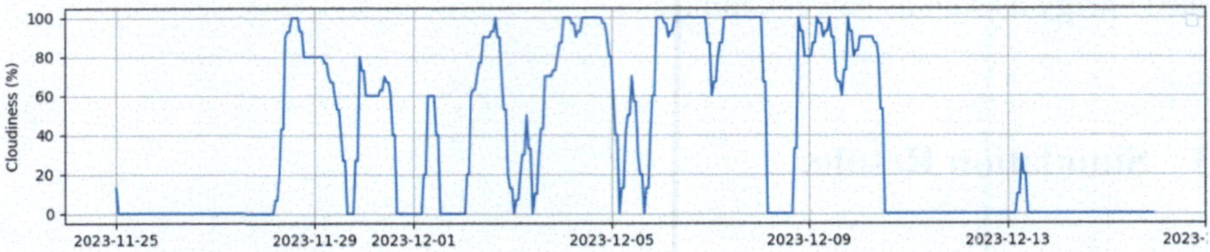

Fig. 6 Cloudiness

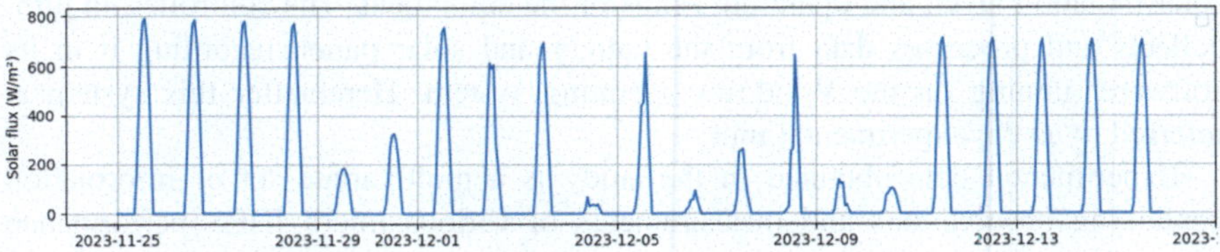

Fig. 7 Solar flux intensity considering cloudiness

Air temperature and humidity. Measurements of ambient air temperature and humidity allow assessing the impact of climatic conditions on IHES operation and solar energy conversion efficiency.

Power parameters. Power generation and consumption data allow assessing the efficiency of a power storage system (battery) and the balance between power generation and consumption.

Equipment condition. Information on equipment condition, including data from sensors and monitoring systems, allows to monitor the performance of IHES components and analyse the causes of possible failures and malfunctions.

The software for visualizing simulated data in plots accordingly performs a number of tasks related to the analysis and output of solar activity and cloudiness data. After all calculations, diagrams reflecting various aspects of solar activity and cloudiness are plotted, and users are given the option of selecting time periods to analyse the data. The diagrams are displayed to the user for evaluation and further analysis. These diagrams are shown in Figs. 5, 6, and 7. And Fig. 8 shows a diagram from IHES monitoring system, which provides an option to compare the data calculated by software with the real data.

As Figs. 7 and 8 show, the data received from the software show similarities with the results obtained from IHES monitoring system. However, there are some

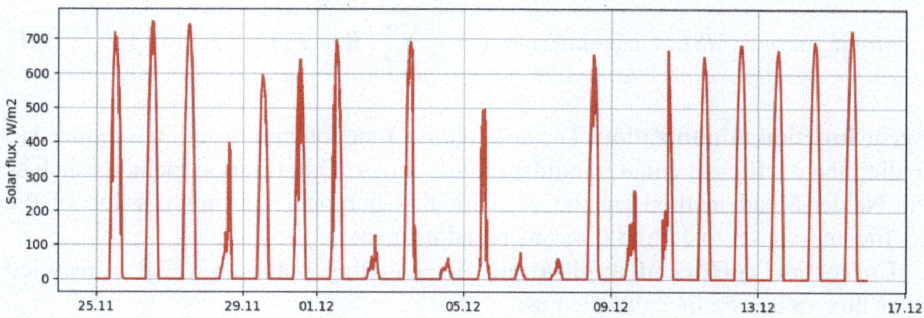

Fig. 8 Solar flux intensity from IHES monitoring system

differences between these data. This may be due to the fact that the data obtained from VentuSky website is not fully reliable due to various factors such as data collection methods, measurement conditions and external influences.

To increase the similarity between simulated and experimental data, it is necessary to calculate a correction coefficient.

The method of sum of squares difference minimisation incorporating regularisation is applied to rectify simulated solar flux data using reference solar flux data from the experimental unit. This method allows finding a correction coefficient that minimises the discrepancy between rectified solar flux data and reference solar radiation data.

Data merging. First, we need to merge two data sets by time. Let $F = \{F_1, F_2, \ldots, F_n\}$ be the measured solar flux data and $R = \{R_1, R_2, \ldots, R_m\}$ be the reference solar radiation data. Data are merged so that the values from F and R correspond to the closest timestamps.

Error function specification. Error function purpose is to minimise the difference between rectified solar flux $\widehat{F}_i$ and measured solar radiation R_i. Mean Squared Error (MSE) is used as the primary measure of error:

$$MSE = \frac{1}{n}\sum_{i=1}^{n}\left(R_i - \widehat{F}_i\right)^2 \tag{49}$$

where $\widehat{F}_i = c \cdot F_i$, and c is the correction coefficient.

Regularization is applied to stabilise the coefficient and prevent it from deviating excessively from unity. Regularisation helps to smooth the correction coefficient and prevent over-learning. Regularization formula:

$$regularization = \lambda \cdot (c-1)^2 \tag{50}$$

where λ is a regularization parameter that determines regularisation's influence on the overall error. Final error function:

$$total\_error = MSE + regularization = \frac{1}{n}\sum_{i=1}^{n}\left(R_i - \widehat{F}_i\right)^2 + \lambda \cdot (c-1)^2 \quad (51)$$

Error function minimization. The task of error function minimisation is reduced to finding the coefficient c that minimises *total _ error*. Optimisation methods such as the Nelder-Mead method can be used for this purpose. The initial value of the coefficient c is set to 1. And 0 means no adjustment.

Correction coefficient application. After finding optimum value c, rectified solar flux values $\widehat{F}_i$ are calculated as:

$$\widehat{F}_i = c \cdot F_i \quad (52)$$

If the initial solar flux F_i and solar radiation R_i data differ significantly, the method of sum of squares minimisation allows finding a coefficient c that minimises the discrepancy between rectified solar flux values and reference radiation values, thus improving the accuracy of measurements.

To determine the power generation of a solar panel, its physical characteristics such as surface area, photocell efficiency and external operating conditions are to be taken into account. Formulas for calculating the effective area of a solar panel and its efficiency factors can be based on geometric and thermodynamic principles as well as manufacturer's data.

System power balance is determined based on the output data of solar radiation, solar panel characteristics, power consumption and battery parameters.

For a battery, the important parameters are its nominal capacity and coefficients that determine the discharge rate and operation mode. Formulas for calculating useful and actual capacity, as well as maximum charge current, can be based on the principles of electrical engineering and batteries specification used [19].

In order to assess the power balance in a system, both the power generation and consumption processes as well as storage and distribution processes are to be taken into account. This optimises system performance and ensures continuous power supply under various operating conditions.

The method for determining battery balance involves several basic steps.

The first step is to calculate time differences between successive measurements. For each timepoint, the difference in hours compared to the previous timepoint is calculated:

$$hour\_diff_i = \frac{datetime_i - datetime_{i-1}}{3600}, i = \overline{1, n}. \quad (53)$$

The next step is to interpolate missing solar flux values. This means that missing values are filled by linear interpolation to ensure continuity of the data.

The third step is to calculate an instantaneous power generated by the solar panel. Using the efficiency of a photocell and solar panel area, the instantaneous power is calculated as:

$$instant\_power_i = solar\_flux_i \cdot solar\_coefficient_i \cdot photoCellSquare \tag{54}$$

The fourth step is to integrate the instantaneous power to calculate the generated power. The power generated is calculated by summing products of instantaneous power and time periods:

$$solar\_energy_i = \sum_{j=0}^{i} instant\_power_j \cdot hour\_diff_j, i = \overline{1,n}, j = \overline{1,m}. \tag{55}$$

The next step is to determine flows in the battery. An energy flow to the battery is determined based on the difference between generated power, taking the charging current limitations:

$$battery\_power_i = \begin{cases} instant\_power_i - load\_power, \\ maxChargeCurrent \cdot charge\_voltage, \end{cases} \tag{56}$$

$$\begin{matrix} if \ \frac{instant\_power_i - load\_power}{charge\_voltage} < maxChargeCurrent \\ if \ \frac{instant\_power_i - load\_power}{charge\_voltage} > maxChargeCurren \end{matrix}, i = \overline{1,n}.$$

The final step is to calculate accumulated power in the battery. Battery power is calculated by taking into account the previous power level and the energy flow into the battery, taking the limits of battery capacity:

$$battery\_energy_i = \begin{cases} batteryCapacity \cdot system\_voltage, \\ 0, \\ prev\_energy + power\_diff, \end{cases} \tag{57}$$

$$if \ prev\_energy + power\_diff \geq batteryCapacity \cdot system\_voltage$$
$$\text{если} \ prev\_energy + power\_diff \leq 0$$
$$if \ prev\_energy + power\_diff < batteryCapacity \cdot system\_voltage$$

The method of power shortage determination as well as the method of power balance determination in the battery includes several basic steps.

It starts by setting an initial level of power shortage. The initial power shortage is defined as the difference between minimum allowable charge level and the battery current power, if it is positive:

$$energy\_shortage_0 = max(0, DoD - battery\_energy_0) \tag{58}$$

The second step is to calculate the power shortage for each time period. That is, for each subsequent time period, the power shortage is determined based on the previous shortage value, power consumption and time period:

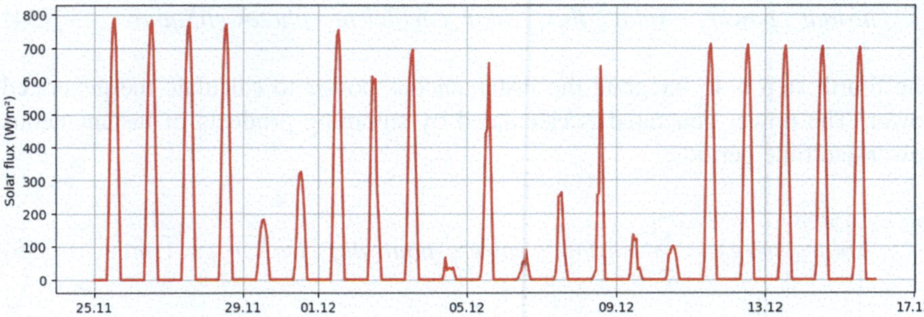

Fig. 9 Insolation diagram based on simulation data

$$energy\_shortage_i = \begin{cases} prev\_de, \text{ если } battery\_energy_i \geq DoD \text{ и } prev\_de = 0 \\ prev\_de + delta\_de, \text{ если } battery\_energy_i < DoD \\ max(0, prev\_de - delta\_de), \text{ если } battery\_energy_i \geq DoD \text{ и } prev\_de > 0 \end{cases} \quad (59)$$

where $delta\_de = load\_power \cdot hour\_diff_i, i = \overline{1, n}$.

The methods described for determining the balance and power deficit in a battery help to analyse the current charge level, identify power shortages and make decisions to optimize system performance.

Based on the data, the solar panel effective area was determined to be 0.431 m^2. For proper system operation, available battery capacity is 9.52 A · h, while the actual battery capacity is 13.6 A · h. The lowest allowable battery capacity for ensuring correct system operation is 4.08 A · h.

Figure 9 shows the insolation diagram for the period from November 25, 2023 to December 15, 2023. Calculations were performed based on simulation data.

Figure 10 illustrates the balance of power accumulated in system's battery plotted using simulated data. It is interesting to note that over the entire tracking period, there were 7 cases when the battery power dropped below the maximum discharge rate depth of 4.08 A · h. Similarly, there were 5 times when the battery was completely discharged to below 0.01 A · h.

To confirm the calculation validity, similar calculation was performed for the experimental unit (EU). Figure 11 shows the insolation diagram, which is based on experimental unit data for the same time period as for the simulation data, from November 25, 2023 to December 15, 2023.

Figure 12 shows the battery power balance diagram for the experimental unit. This diagram shows the change in a battery power level over the considered time period. The diagram shows that during 7 time periods, the battery power was below the maximum discharge rate depth, which is 4.08 A · h. It is also noted that for 3 time periods the battery was completely discharged, i.e. its power level dropped below 0.01 A · h.

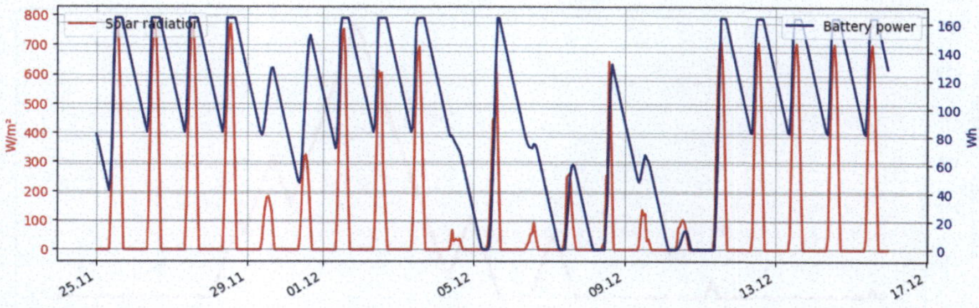

Fig. 10 Battery power balance based on simulated data: blue line means value of accumulated power, red line means value of insolation

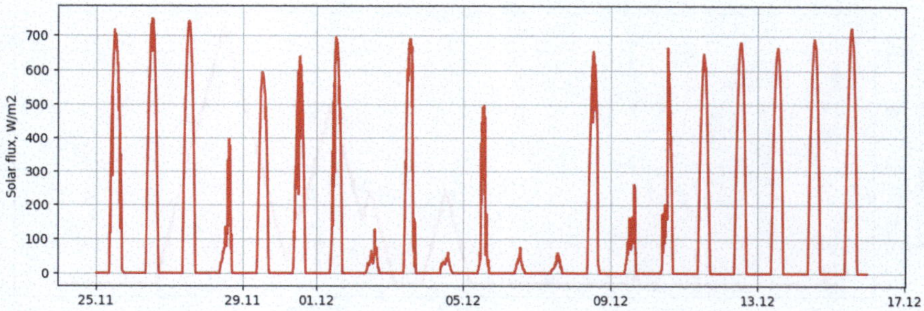

Fig. 11 Insolation diagram based on data from EU

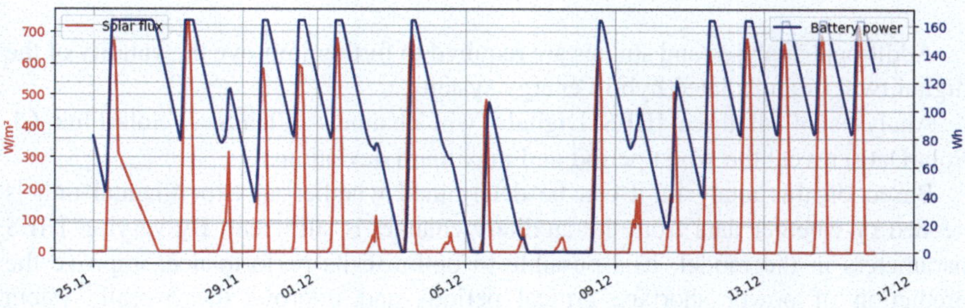

Fig. 12 Battery power balance diagram based on EU data: blue line means value of accumulated power, red line means value of insolation

It is also important to note that analysing the two Battery Power Balance diagrams for simulated data and the data from the experimental unit reveals differences in their characteristics. There is a difference in the number of time periods when the battery power level reaches critically low values. Specifically, the simulated data has 5 time periods when the battery was completely discharged, i.e. the power level dropped below 0.01 $A \cdot h$. While 3 such periods are observed in the data from the experimental unit. Figures 13 and 14 show power shortage diagrams of EU and IHES DT.

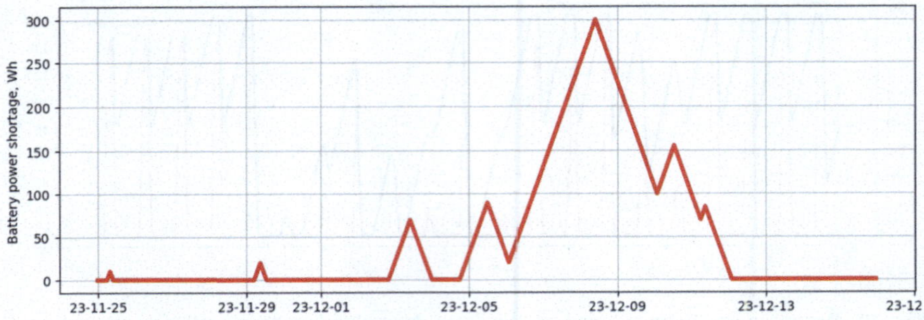

Fig. 13 EU power shortage diagram

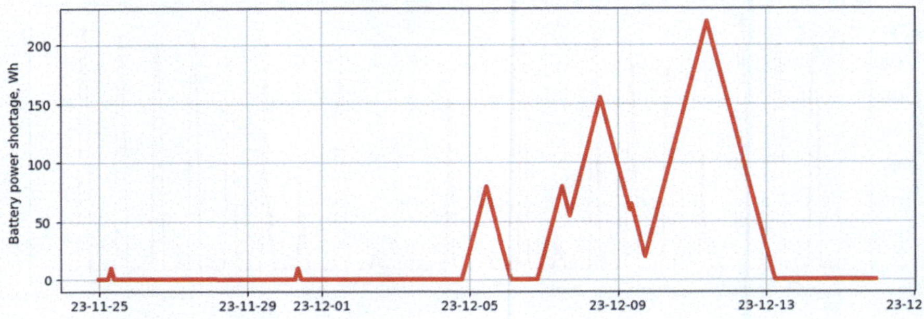

Fig. 14 IHES DT power shortage diagram

Additional analyses and studies are required to further improve the validity of the digital twin of an isolated hybrid energy system.

Analyses of all three IHES Digital Twin channels (ClearSky, SolarFlux_Cl, SolarData) reveal the worst period and maximum deviations.

Based on this analysis, it can be determined whether an experimental unit is needed or whether data from the auxiliary channel is sufficient. By varying IHES parameters in the model, it is possible to optimise its performance, improve the prediction of power shortage critical periods and improve the overall system efficiency.

The method for calculating the correction coefficient is required to rectify solar flux data calculated by IHES DT.

The method for calculating the correction coefficient using sum-of-squares difference minimisation includes regularisation, which allows the measured solar flux to be rectified using reference solar radiation data. Error function purpose is to minimise the difference between rectified solar flux (IHES DT) and measured solar radiation (EU) using Mean Squared Error (MSE) as the main error metric. Regularisation helps to avoid the coefficient deviating too much from unity. The Nelder-Mead minimisation method is used to find the correction coefficient that minimises the error function. Correction result is shown in Fig. 15.

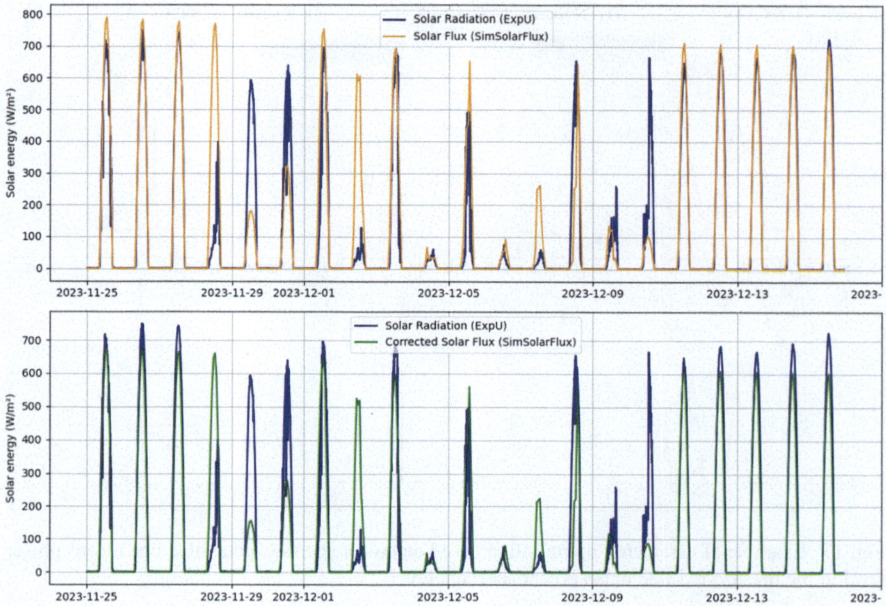

Fig. 15 Correction results: upper diagram shows data before correction (blue line means radiation level from the experimental unit, orange line means simulation data before correction); lower diagram shows data after correction (blue line means radiation level from the experimental unit, green line means simulation data after correction)

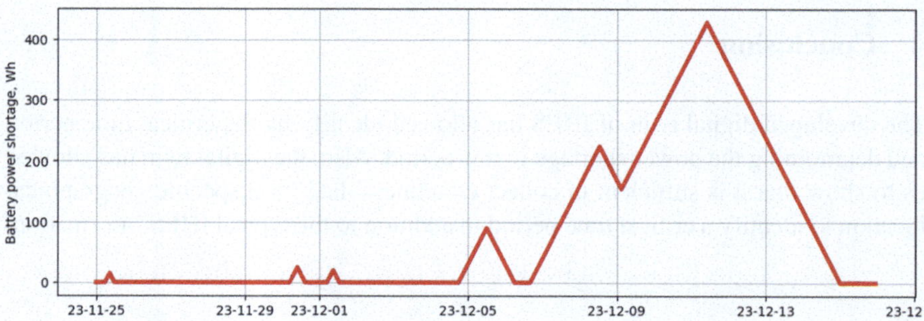

Fig. 16 Power shortage of IHES DT after correction

The values of power shortage are expected to change after rectifying solar energy data of IHES DT. These values are presented as a diagram in Fig. 16.

Analyses of all three IHES Digital Twin channels (ClearSky, SolarFlux_Cl, SolarData) reveal the worst period and maximum deviations (Fig. 17).

The top row in Fig. 17 shows the power shortage when successively increasing the number of batteries from one to three, while keeping one solar panel. In the bottom row, the number of batteries is similarly increased, but the number of panels

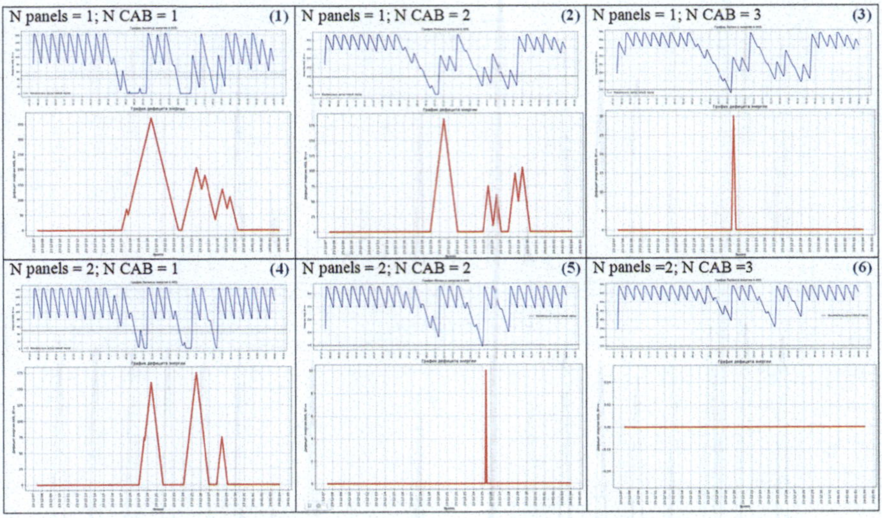

Fig. 17 Example of parametric optimisation based on power shortage data: blue line means power level in the battery, red line mean power shortage level

is two. The last part (lower right corner) shows that with two solar panels and three batteries there is no power shortage.

4 Conclusion

The developed digital twin of IHES has allowed identifying the critical time period and determining the power shortage in this period. Also, the digital twin has allowed us to show that it is sufficient to collect cloudiness data for a specific geographical location to identify a critical time period in addition to the typical IHES information.

References

1. Yu, S., You, L., Zhou, S.: A review of optimization modeling and solution methods in renewable energy systems. Front. Eng. Manag. **10**, 640–671 (2023)
2. Qiu, K., Entchev, E.: Modeling, design and optimization of integrated renewable energy systems for electrification in remote communities. Sustain. Energy Res. **11**, 10 (2024)
3. Adeyinka, A.M., Esan, O.C., Ijaola, A.O., et al.: Advancements in hybrid energy storage systems for enhancing renewable energy-to-grid integration. Sustain. Energy Res. **11**, 26 (2024)
4. Abdelsattar, M., Mesalam, A., Diab, A.A.Z., et al.: Optimal sizing of a proposed stand-alone hybrid energy system in a remote region of Southwest Egypt applying different meta-heuristic algorithms. Neural Comput. Applic. **36**, 16251–16269 (2024)

5. Peng, Y., Zhang, Y.: Energy optimization of stand-alone electrical grid considering the optimal performance of the hydrogen storage system and consumers. J. Eng. Appl. Sci. **71**, 61 (2024)
6. El Boujdaini, L., Mezrhab, A., Moussaoui, M.A., et al.: Sizing of a stand-alone PV–wind–battery–diesel hybrid energy system and optimal combination using a particle swarm optimization algorithm. Electr. Eng. **104**, 3339–3359 (2022)
7. Amoussou, I., Paddy, E.Y., Agajie, T.F., et al.: Enhancing residential energy access with optimized stand-alone hybrid solar-diesel-battery systems in Buea, Cameroon. Sci. Rep. **14**, 15543 (2024)
8. Mundu, M.M., Nnamchi, S.N., Sempewo, J.I., et al.: Simulation modeling for energy systems analysis: a critical review. Energy Inform. **7**, 75 (2024)
9. Mitrofanov, S.V., Perepelkin, K.A.S.V.: Mathematical modelling of hybrid wind-solar station for power supply of own needs Series: Energy, pp. 18–26. Vestnik of the South Ural State University (2022)
10. Gurumoorthi, G., Senthilkumar, S., Karthikeyan, G., et al.: A hybrid deep learning approach to solve optimal power flow problem in hybrid renewable energy systems. Sci. Rep. **14**, 19377 (2024)
11. Altshuller, G.: 40 principles: TRIZ keys to technical innovation, p. 1. Technical Innovation Center, Inc (2002)
12. Kovalev, S.P.: Information support design for power systems digital twins, pp. 66–81. Systems and Tools in Computer Science (2020)
13. Massel, L.V., Massel, A.G.: Semantic modelling in digital twin design of energy objects and systems. Des. Ontol. **13**(1), 44–54 (2023)
14. Bayer, D.A., Bayer, M.P.: Digital twin of a local energy system based on real smart meter data. Energy Inform. **6**(1), 8 (2023)
15. Kovalev, S.P.: Application of deep learning neural networks in mathematical support of electrical power systems digital twins, pp. 133–144. Systems and Tools in Computer Science (2021)
16. Kandemir, E., Hasan, A., Kvamsdal, T., et al.: Predictive digital twin for wind energy systems: a literature review. Energy Inform. **7**, 68 (2024)
17. Sleiti, A.K., Kapat, J.S., Vesely, L.: Digital twin in energy industry: proposed robust digital twin for power plant and other complex capital-intensive large engineering systems. Energy Rep. **8**, 3704–3726 (2022)
18. Yu, W., et al.: Energy digital twin technology for industrial energy management: classification, challenges and future. Renew. Sust. Energ. Rev. **161**, 112407 (2022)
19. Vissarionov, V.I., Derbgina, G.V., Kuznetsova, V.A., Malinin, N.K.: Solar energy, vol. 320. Moscow Power Engineering Institute (2008)

Quadratic Programming in a Model of Resource Distribution in Agriculture Based on Quantum Approximate Optimization Algorithm

L. U. Safarova, D. T. Muhamediyeva, and D. Vasiyeva

Abstract This paper considers the application of quantum algorithms to the problem of resource allocation optimization in agriculture. In this paper, the approach of applying the quantum algorithm Quantum Approximate Optimization Algorithm to the problem of resource allocation optimization in agriculture was investigated. The paper develops an algorithm for applying Quantum Approximate Optimization Algorithm to the problem of resource allocation optimization. This algorithm includes creating a quantum chain representing the optimization problem, applying evolution operators and measurements to find the state corresponding to the optimal solution to the problem. The application of Quantum Approximate Optimization Algorithm has proven its effectiveness in solving complex optimization problems in agriculture. Quantum Approximate Optimization Algorithm is able to take into account many factors, such as resource availability, crop requirements and other constraints, which allows obtaining optimal solutions to the resource allocation problem.

Keywords Quantum algorithms · Quantum approximate optimization algorithm · Optimization · Resource allocation · Agriculture · Mathematical model · Combinatorial optimization · Production efficiency · Agricultural sector

1 Introduction

1.1 A Subsection Sample

Optimizing the allocation of resources in agriculture plays a key role in improving production efficiency, increasing output and improving the sustainability of the agricultural sector. The complexity of this task lies in the need to take into account

L. U. Safarova (✉) · D. T. Muhamediyeva · D. Vasiyeva
Tashkent Institute of Irrigation and Agricultural Mechanization Engineers National Research University, Tashkent, Uzbekistan

A. Gibadullin, S. Gordin (eds.), *Computing Technologies and Applied Mathematics*, Springer Proceedings in Mathematics & Statistics 500,
https://doi.org/10.1007/978-3-031-90546-9_2

many factors, such as resource availability, crop requirements, climatic conditions, market requirements, etc.

Traditional optimization methods, such as linear or quadratic programming, may face limitations in the effectiveness of solving a problem due to its complexity and dimensionality. In recent years, interest in the application of quantum algorithms in optimization problems has been growing. Quantum Approximate Optimization Algorithm (QAOA) is one of the promising quantum algorithms designed to solve combinatorial optimization problems. The application of QAOA to the task of resource allocation in agriculture represents a new approach that can lead to more efficient and sustainable use of resources in the agricultural sector. This paper explores the possibility of applying QAOA to the task of optimizing resource allocation in agriculture. The basic principles of the QAOA operation will be considered, as well as the specifics of its application to optimization problems in agriculture. It is assumed that the use of a quantum algorithm can lead to more accurate and faster solutions to the optimization problem and contribute to improving the efficiency of agricultural production. The use of quantum algorithms in optimization problems, such as resource allocation in agriculture, is an active area of research. One of the quantum algorithms that can be applied to such tasks is the Quantum Approximate Optimization Algorithm (QAOA) [1–4].

To apply QAOA to the problem of resource allocation in agriculture, you first need to present the problem as an optimization problem that can be solved using QAOA. Then quantum optimization methods are applied to solve this problem. The specific QAOA algorithm involves the creation of a quantum circuit that represents an optimization problem in the form of a Schrodinger operator. Then the evolution and measurement operators are applied to find the state corresponding to the optimal solution of the problem. The application of QAOA to the task of resource allocation in agriculture requires additional work to adapt the algorithm to the specific conditions and requirements of this task. This may include the development of special cost functions, taking into account resource constraints, etc. In general, the use of quantum algorithms such as QAOA to solve problems of resource allocation in agriculture can offer new optimization methods and allow finding more effective solutions in comparison with classical optimization methods [5–7].

The scientific novelty of this work lies in the application of the Quantum Approximate Optimization Algorithm (QAOA) to the problem of optimizing resource allocation in agriculture. Currently, quantum algorithms are actively being investigated in the context of solving combinatorial optimization problems, but their application to optimization problems in agriculture is relatively new. The application of QAOA to the task of resource allocation in agriculture represents a new approach that can lead to more efficient use of resources and improved production in the agricultural sector. This research introduces quantum technologies into the field of agriculture, opening up new perspectives for solving complex optimization problems in this industry. It should also be noted that the application of quantum algorithms to optimization problems in agriculture requires taking into account the specific features of this industry, such as seasonal changes, climatic conditions, market requirements and others. Therefore, this study also includes the adaptation of QAOA to specific agricultural conditions, which is of additional scientific value [8–10].

2 Materials and Methods

Let's consider nonlinear models in the presence of constraints [1]:

$$F(\overline{x}) \to \min, \overline{x} \in E^n, \quad (1)$$

$$h_i(\overline{x}) = 0; i = \overline{1, m}, \quad (2)$$

$$g_i(\overline{x}) \geq 0; i = \overline{m+1, p} \quad (3)$$

Nonlinear programming methods are widely used to solve such problems. One approach is to reduce a nonlinear problem to a linear programming problem. To do this, a linearization procedure is used, which converts nonlinear functions into linear approximations in the vicinity of a certain point. This then allows you to apply methods for solving linear programming problems to find the optimal solution.

In this case, the problem of nonlinear modeling in the general formulation can be modified by writing each of the nonlinear functions of this problem with the first two numbers in the corresponding Taylor series expansion in the neighborhood and reduced to a linear programming problem [1, 2]:

$$\{f\left(\overline{x}^{(k)}\right) + \nabla^r h_i\left(\overline{x}^{(k)}\right)\left(\overline{x} - \overline{x}^{(k)}\right)\} \Rightarrow \min, \overline{x} \in E^n \quad (4)$$

under restrictions:

$$h_i\left(\overline{x}^{(k)}\right) + \nabla^T h_i\left(\overline{x}^{(k)}\right)\left(\overline{x} - \overline{x}^{(k)}\right) = 0, i = \overline{1, m} \quad (5)$$

$$g_i\left(\overline{x}^{(k)}\right) + \nabla^T g_i\left(\overline{x}^{(k)}\right)\left(\overline{x} - \overline{x}^{(k)}\right) \geq 0, i = \overline{m+1, p} \quad (6)$$

The linearization procedure is often used in an iterative process. It allows you to linearize nonlinear functions at the current point at each iteration and solve the corresponding linear problem. The iteration process then continues until a stop condition is reached, such as convergence to an optimal solution or reaching the maximum number of iterations. This approach is widely used in control problems, where nonlinear models can be approximated to linear problems.

The convergence of linear approximation methods to the solution is guaranteed when the conditions are met:

Functions $f(\overline{x}), h_1(\overline{x}), \ldots, h_m(\overline{x}), g_{m+1}(\overline{x}), \ldots, g_p(\overline{x})$ they must be continuous and differentiable. Function $f(\overline{x})$ It should be convex, and the restrictions $\sum h_j^2(\overline{x})$. They must be convex in an acceptable area.

Functions $g_{m+1}(\overline{x}), \ldots, g_p(\overline{x})$they must be concave.

The area D must be closed, convex, and non-empty.

All parameters found in the function conditions must be limited.

If these conditions are not met, methods using linear approximation can only find a local optimum.

In the context of linear programming (LP) tasks, an additional condition is added to coordinate decisions and take into account assumptions: limiting the step length when moving in different directions to a small value $\delta_j^{(k)} > 0$. This helps to prevent steps that are too large, which can lead to an exit from the area where the approximation is valid:

$$\delta_j^{(k)} - \left|\overline{x}_j^{(k+1)} - \overline{x}_j^{(k)}\right| \geq 0, j = \overline{1, n} \tag{7}$$

Solving the problem (4)–(6) allows you to calculate the next point:

$$\overline{x}^{(k+1)} = \overline{x}^{(k)} + F^0\left(\widehat{x}, \overline{x}^{(k)}, k\right) \tag{8}$$

Re-applying this procedure with a lower value $\delta^{(k)}$ this happens until the minimizing correction to the value found in the previous step is less than the preset number. This allows you to get a more accurate solution to the problem.

If the initial vector is $\overline{x}^{(0)}$ it turns out to be outside the permissible range, artificial variables are introduced $\omega_i^{(k)} > 0$, which are added to the constraints of equality and inequality. This allows you to bring the starting point to an acceptable area and start the optimization process.

The generalized reduced gradient method, which is included in the class of optimization algorithms and is used in [1, 5]. This method is effective in solving problems with a continuous and differentiable goal function based on control variables, as well as in the case when the task parameters are in continuous ranges. The generalized reduced gradient method is an iterative optimization method that uses information about the gradient of the objective function to find the optimal solution. It combines the ideas of gradient descent with the method of finding the optimal step, which makes it effective in solving various optimization problems. However, it is worth noting that this method is not always applicable in cases where the goal function is not continuous and differentiable, or when the task parameters have discrete values or are in discrete ranges. In such cases, other optimization methods specifically adapted to these conditions may be required.

When the control or coordinating parameters have discrete values, other optimization methods are used. One of these methods is dynamic programming, which is widely used to solve optimization problems in conditions of discrete parameters.

The decomposition of the problem allows you to use the method that is most effective for this situation in solving each local problem. Parametric decomposition involves fixing some parameters and solving the main problem taking into account these fixed parameters. This allows you to simplify the task and focus on optimizing specific parameters. Parametric decomposition is an approach to solving complex optimization problems, which consists in dividing the main task into simpler subtasks and then combining their results. The main idea is to fix some parameters of the problem and solve it under the conditions of these fixed parameters. First, the parameters that will be fixed during the decomposition process are determined. The choice of these parameters may depend on the specific task and optimization goals. The task is divided into smaller subtasks, taking into account fixed parameters. This can be done in various ways, such as time partitioning, spatial partitioning, or any other approach that takes into account the specifics of the task. Each subtask is solved independently, taking into account fixed parameters. Various optimization methods can be used to solve these problems, such as dynamic programming for discrete parameters or optimization methods for continuous parameters. After solving all local problems and obtaining optimal values for each subtask, the results are combined to obtain a solution to the main problem. Parametric decomposition makes it possible to effectively solve complex optimization problems by breaking them down into simpler components. This approach is especially useful when the control or coordinating parameters have discrete values, and can be effectively applied in a wide range of applied tasks.

With parametric decomposition [1, 4], r parameters are fixed $(x_{n+1}, \ldots, x_{n+r})$ and the main task is being solved (1):

$$\begin{aligned}
&\sum_{j=1}^{n} a_{ij}x_j + g_i(x_{n+1}, \ldots, x_{n+r}) = b_i, i = \overline{1,m} \\
&\underline{U_k} \le x_{n+k} \le \overline{U}_k, k = \overline{1,r} \\
&x_j \ge 0, j = \overline{1,n}
\end{aligned} \tag{9}$$

$$F = \sum_{j=1}^{n} a_{0j}x_j + g_0(x_{n+1}, \ldots, x_{n+r}) \to \min \tag{10}$$

the dimension of which is reduced by r [1]. As a result of solving this linear programming problem, basic, non-basic variables and dual estimates are found Π_1, …, Π_m for constraints (9). After that, the second problem is solved:

$$\sum_{i=1}^{n} \Pi_i g_i(x_{n+1}, \ldots, x_{n+r}) + g_0(x_{n+1}, \ldots, x_{n+r}) \to \min, \tag{11}$$

under restrictions:

$$\overline{x}^{(k+1)} = \overline{x}^{(k)} + F^0\left(\widehat{x}, \overline{x}^{(k)}, k\right),$$
$$\underline{U_k} \le x_{n+k} \le \overline{U}_k, k = \overline{1, r}. \tag{12}$$

We find all non-basic variables in problem (12) that were found when solving problem (9) and (10), and equate them to zero. We perform a linear approximation of constraints (12) and the objective function in order to transform the original nonlinear problem into a linear problem. We solve the resulting linear problem. The resulting solution is used to find a new solution to the original problem (9) and (10). The procedure is repeated until the specified accuracy of the solution is reached. This process makes it possible to solve complex nonlinear optimization problems by linearization and subsequent iterative optimization. Carrying out the structural decomposition of the initial problem by linearization of the initial model opens up new possibilities for using linear models with variable coefficients. This makes it possible to effectively solve a variety of optimization problems, since linear models often have well-developed solution methods and can be solved with high accuracy and efficiency.

The use of these models makes it possible to solve a problem of the following type [1, 9]:

$$G_i = \sum_{j=1}^{n} f_{ij}(\overline{u})x_j - b_i = 0, i = \overline{1, n};$$
$$x_j \ge 0, j = \overline{1, R}; \tag{13}$$

$$F = \sum_{j=1}^{n} F_j(\overline{u})x_j \to \min \tag{14}$$

where f_{ij}, F_j—set control functions $\underline{u} = \{u_1, \ldots, u_t\}, u \in U$.

Such a model can be written when the operation of each element is described by a linear model with variable coefficients of the form:

$$y_i^{(k)} = \sum_{j=1}^{R_i} f_{ij}^{(k)}\left(\overline{u}^{(k)}\right)x_j^{(k)}, i = \overline{1, m}$$
$$x_j^{(k)} \ge 0, j = \overline{1, R_i} \tag{15}$$

$$F^{(k)} = \sum_{j=1}^{R_i} F_j^{(k)}\left(\overline{u}^{(k)}\right)x_j^{(k)} \to \min \tag{16}$$

The model of an individual element (aggregate) can be considered as the result of the linearization of a nonlinear mode:

$$y_i^{(k)} = f_i^{(k)}\left(\overline{x}^{(k)}, \overline{u}^{(k)}\right), i = \overline{1, m_k} \tag{17}$$

by input parameters$\overline{x}^{(k)}$, for example, by decomposing a function $f_i^{(k)}\left(\overline{x}^{(k)}, \overline{u}^{(k)}\right)$ in the Taylor series in the vicinity of a certain value of the input parameters $\overline{x}_0^{(k)}$ and neglecting in this decomposition the terms of the second order and above [1–3].

The concept of a boundary mode is used to approximate the production capabilities of units that can operate in various modes within a time interval.

$[\underline{u}_i, \overline{u}_i]$. The boundary conditions represent the maximum possible operating conditions of technological units, which are determined for each unit. Intermediate modes of operation of the units can be approximated using linear combinations of boundary modes. This allows you to take into account various options for the operation of the units within a given time interval. By sequentially fixing the parameter values in the interval $[\underline{u}_i, \overline{u}_i]$. For each of these values, using the principle of parametric decomposition, it is possible to obtain a sequence of optimal values of the functional *F*. Then, from this sequence, a solution is selected that gives the lowest value of the functional *F*.

Thus, the principle of parametric decomposition allows us to consider various options for the operation of aggregates within a given time interval and choose the optimal solution taking into account all possible options. This approach is an effective way to take into account a variety of conditions and constraints when optimizing production processes.

If an object has the property of internal optimization and seeks to maximize its usefulness or other functionality in changing conditions, then this can contribute to self-regulation of its behavior. This property can allow an object to choose sustainable and rational strategies in response to changes in resource provision and expense ratios. If an object is able to independently analyze and evaluate the environment, as well as its internal state, it can adapt to changing conditions and choose optimal actions without direct external control. However, it should be borne in mind that rapid and inaccurate changes in conditions, as well as restrictions on access to information, can complicate the process of self-regulation and decision-making. To solve such problems, non-linear programming methods can be used, which allow taking into account complex relationships and limitations in the system. Modeling the behavior of an object and optimizing its strategies can help you understand how an object will respond to various scenarios of changes in the environment.

This nonlinear programming problem is written as follows [1, 4–7]:

Let x be a vector of products and R be a vector of resources. We have an additive utility function:

$$f(x, \alpha)$$

$$y=f(x,\alpha)=\sum_{j=1}^{n}\alpha_j f_j(x_j)\rightarrow \max \tag{18}$$

and additive resource constraints g(x):

$$g(x)=\sum_{j=1}^{n}g_j(x_j,\beta_j)\leq R, \tag{19}$$

$$x\in Q. \tag{20}$$

where j = 1,…,n—product index; $R = (R_1, \ldots, R_m)$—resource vector; i = 1,…,m—resource index; $f_j(x_j)$. The utility function of production j—of the product x_j; $f(x, \alpha)$—the utility function of the total volume j- x products; $g_j(x_j, \beta_j)$—a vector strictly convex function of resource consumption in production j—of the product x_j; β_j—parameter vector j—th technologies, $\beta_j \in B_j$convex sets; Q—convex set.

Let x^*—solving the problem (18)–(20), a $f(x, \alpha)$. It is measured as the total amount of material goods.

The model should take into account the uncertainty in the parameter values, and not just accept them as fixed numbers. This approach allows for a more adequate reflection of real conditions and consideration of possible variations in parameters, which is important for the accuracy and reliability of the model. Modeling the uncertainty of parameters in optimization problems leads to the concept of stochastic programming or programming with uncertain data. In such models, the parameters can take values from a certain probability distribution, and the goal is to find the optimal management strategy under uncertain conditions. The use of stochastic programming allows you to take into account uncertainty and risks when making decisions, which is especially important in a variable environment and various factors affecting production processes. Uncertainty modeling can also include various risk analysis techniques such as sensitivity analysis, scenario analysis, and probabilistic analysis to assess the possible consequences of different parameter value options.

This approach can help to create more flexible and adaptive management strategies that can work effectively in conditions of uncertainty and variability of the external environment [1, 11].

Let's write out the Lagrangian of the problem (18)–(20):

$$L=\sum_{j=1}^{n}\alpha_j f(x_j)-\sum_{j=1}^{n}\lambda_i\left[R-g_j(x_j,\beta_j)\right]^2. \tag{21}$$

We have a system of equations:

$$\begin{cases} \dfrac{\partial L}{\partial x_j} = \alpha_j \dfrac{\partial f}{\partial x_j} - 2\lambda_j \dfrac{\partial g_j}{\partial x_j} = 0, \\ \dfrac{\partial L}{\partial \lambda} = R - g_j(x_j, \beta_j) = 0. \end{cases} \tag{22}$$

Any solution of a system of equations determines the point at which the extremum of the function can take place $f(x, \alpha)$. Therefore, by solving the system of Eq. (22), all points are obtained at which the function (18) can have extreme values. Among the points that are suspicious of the extremum, there are such x^*, at which the extremum is reached, and the values of the function (18) are calculated at these points.

3 Results

The problem of resource allocation in agriculture is considered as the initial optimization problem. This includes the definition of a target function (profit maximization), restrictions on the use of resources (linear and/or quadratic) and other factors characterizing production processes in agriculture. This study uses quantum optimization algorithms, QAOA (Quantum Approximate Optimization Algorithm), to solve the quantum optimization subtask that is included in the ADMM optimization process.

Based on the data obtained, a mathematical model of the resource allocation optimization problem is being developed. The model takes into account the objective function (profit maximization), restrictions on the use of resources and other factors characterizing production processes. To account for quadratic programming in the agricultural resource allocation model, we can change the objective function so that it takes into account quadratic dependencies between variables. Let's assume that we have a quadratic relationship between the amount of resources used and profit, that is, we strive to maximize the quadratic function of profit relative to the use of resources.

Let Q be a quadratic matrix representing the quadratic dependencies between the variables (products) and (resources). Then the objective function will have the form:

$$\sum_{j=1}^{n} c_j x_j - \frac{1}{2} \sum_{i=1}^{m} \sum_{j=1}^{n} Q_{ij} x_i x_j \to \max$$

This function maximizes profit by subtracting the resource usage penalty from it, taking into account the quadratic dependencies between resource usage and profit.

At the same time, the restrictions on the use of resources will remain linear, as described earlier:

$$\sum_{j=1}^{n} a_{ij}x_j \le b_i, \ i = 1, 2, \ldots, m$$

An algorithm for applying QAOA to the task of optimizing resource allocation is being developed. This algorithm includes the creation of a quantum circuit representing an optimization problem, the use of evolution and measurement operators to find a state corresponding to the optimal solution of the problem [11, 12].

1. Setting the initial parameters of the algorithm. The initial parameters of the algorithm, such as the depth of the quantum chain and the initial values of the angles γ and β, they can be selected based on empirical data and the researcher's experience. It is important to take into account that the optimal values of these parameters may depend on the specific optimization task and the properties of the quantum equipment used. The depth of the chain determines the number of layers in the QAOA. Usually, the initial value is selected in the range from 1 to 5. Large values can increase the accuracy of the solution, but at the same time the computational complexity increases. The initial values of the angles can be selected randomly within a given range or using optimization techniques such as random search or gradient-based optimization techniques. These angles are usually selected from the range. After selecting the initial parameters, the QAOA algorithm is executed to generate a solution to the optimization problem. The results obtained can then be evaluated and, if necessary, the parameters can be further adjusted to improve the quality of the solution [13, 14].
2. Creating the initial quantum state $|\psi(\gamma, \beta)\rangle$ It is an important step in the QAOA algorithm and may depend on the specific optimization task. Usually, this initial state is chosen in such a way that it is a superposition of various basic states in order to ensure uniform coverage of the state space and increase the probability of finding an optimal solution.

   As an example, for the task of optimizing resource allocation in agriculture, the initial quantum state can be formed by a superposition of all possible states representing various resource allocation options. This can be realized by applying the Hadamard operator to all qubits, which will lead to the creation of an equally probable superposition of all basic states.

   Thus, the initial quantum state $|\psi(\gamma, \beta)\rangle$ it can be represented as follows [15]:

   $$|\psi(\gamma, \beta)\rangle = H^{\otimes n}|0\rangle^{\otimes n},$$

   where H—the Hadamard operator, n—number of qubits, a $|0\rangle$—the basic state of the qubit.

   This initial superposition state allows us to explore a wide range of possible solutions to the optimization problem in the QAOA algorithm.
3. Application of the evolution operator $U(\gamma, \beta)$. The evolution operator plays a key role in the QAOA algorithm. This operator depends on the optimization task and

on the current values of the parameters. It is usually represented as a sequential application of gates, described as follows:

$$U(\gamma,\beta)=e^{-i\beta_p B}e^{-i\gamma_p C}\ldots e^{-i\beta_1 B}e^{-i\gamma_1 C}$$

where B_i and C_i—these are the Hamiltonians corresponding to the optimization problem, and are the current values of the parameters at the step.

Two types of Hamiltonians are commonly used in the QAOA algorithm: Hamiltonians.

C_i They are Hamiltonians associated with the objective function of the optimization problem.

Hamiltonians

B_i They are Hamiltonians that help rotate the states of qubits around the Schrodinger axis and increase the probability of detecting an optimal solution. They are usually chosen as the sum over all the qubits of the Pauli operators X or Z.

After applying the evolution operator $U(\gamma,\beta)$ to the initial state $|\psi(\gamma,\beta)\rangle$, a new state is obtained. This state contains information about the current solution of the optimization problem and will be used to calculate the objective function and further optimize the parameters γ_i and β_i in the next iteration of the algorithm [16].

4. Steps 2 and 3 of the QAOA algorithm are usually repeated several times with different parameter values

   γ_i and β_i, forming an iterative process. This allows us to gradually improve the quality of solving the optimization problem and approach the optimal solution. The process is repeated until a stop criterion is reached, such as a set number of iterations or achieving a certain accuracy of the solution.
5. After completing the iterations of the QAOA algorithm, the quantum state is measured, which contains information about the current solution to the optimization problem. The measurement leads to the collapse of the quantum state into a classical configuration in which each qubit takes on a certain value. Then, based on the measurement results, the values of the objective function and other characteristics of the optimization problem are estimated. The measurement results obtained are strings of bits that correspond to various possible solutions to the optimization problem. For each measured solution, the value of the objective function is calculated, which allows you to evaluate the quality of the solution. For example, in the task of maximizing profit, this may be the total profit from the allocation of resources, and in the task of minimizing costs, total costs or losses. From all the measured solutions, the one that corresponds to the optimal value of the objective function is selected. This solution is a solution to the optimization problem obtained using the QAOA algorithm. The obtained optimal solution and its value of the objective function are analyzed to understand the effectiveness of the algorithm and its suitability for solving a specific optimization problem. Measuring the quantum state and estimating the values of the optimization problem based on the measurement results allow us to obtain the

final result of the QAOA algorithm and evaluate its effectiveness in solving optimization problems [17].

6. Algorithm parameters

   γ and β They are updated according to the results obtained after each iteration of the QAOA algorithm is completed. This allows you to refine the values of the parameters and improve the quality of solving the optimization problem. The process of updating parameters can be carried out in various ways, depending on the chosen optimization strategy.

7. The process continues until the stop criterion is reached, which is determined in advance and may vary depending on the specific task and user requirements. The algorithm can be stopped after completing a certain number of iterations. This allows you to control the execution time of the algorithm and limit its computational complexity. The algorithm can be stopped when a solution has been reached that satisfies certain quality criteria. For example, the value of the objective function may be close to optimal or satisfy other specified conditions.

   Operator $e^{-i\beta_p B}$ It can be implemented as a quantum gate using single-qubit rotations. The shape of the turn gate $R_z(\theta)$ in the form of a matrix:

$$R_z(\theta) = \begin{pmatrix} e^{-i\theta/2} & 0 \\ 0 & e^{i\theta/2} \end{pmatrix}$$

Thus, the evolution operator $e^{-i\beta_p B}$ it can be implemented as follows:

$$e^{-i\beta_p B} = R_z(-2\beta_p B)$$

This gate $R_z(-2\beta_p B)$ acts on a single qubit and rotates it around the axis Z by an angle proportional to the parameter β_p and weight B.

Let's give an example.
Maximize
−2*x0*x2 − x1*x2 − x0 + 2*x1 − 3*x2
Subject to
Linear constraints (1)
x0 + x1 + x2 < = 1 'cons1'
Binary variables (3)
x0 x1 x2
objective function value: 6.0
variable values: x0 = 1.0, x1 = 0.0, x2 = 1.0
status: SUCCESS. The quantum scheme for solving the problem is shown in Fig. 1.

The Quantum Approximate Optimization Algorithm has been successfully applied to the task of optimizing resource allocation in agriculture. QAOA allowed us to solve the optimization problem, taking into account the complex dependencies

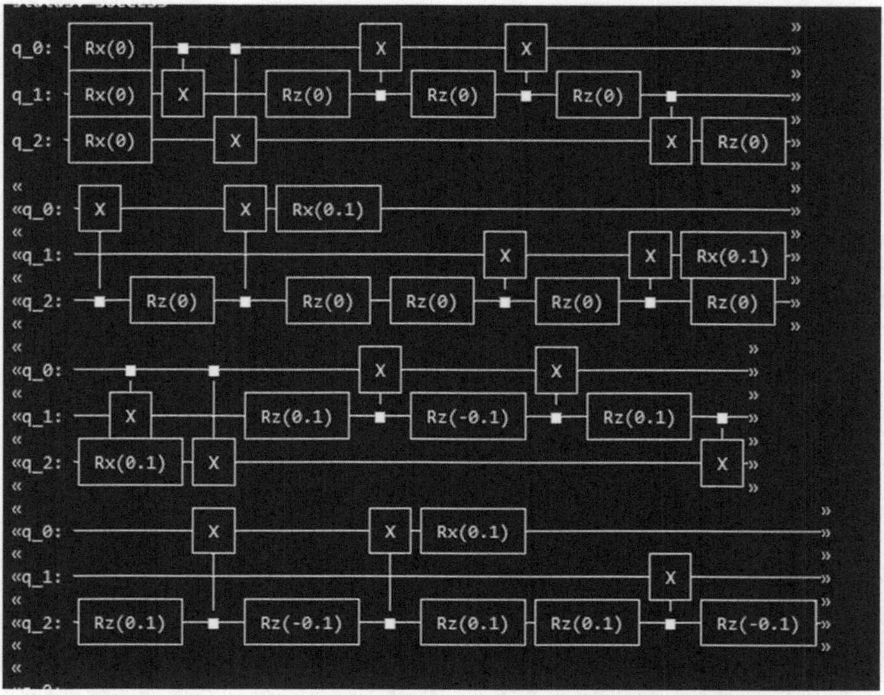

Fig. 1 Quantum scheme for solving the problem

between resources, crops and other factors of production. Optimal solutions to the optimization problem have been obtained, which allow maximizing profits from agricultural production with limited resources. Solutions take into account various factors such as resource availability, crop resource requirements, and other constraints.

4 Conclusion

In this paper, the approach of applying the Quantum Approximate Optimization Algorithm to the task of optimizing resource allocation in agriculture was investigated. The application of QAOA has shown its effectiveness in solving complex optimization problems in agriculture. QAOA is able to take into account many factors such as resource availability, crop requirements, and other constraints, which allows for optimal solutions to the resource allocation problem. The solutions obtained in the application of QAOA demonstrate high quality and efficiency. They allow you to maximize profits from the production of agricultural products with limited resources. The results of the study indicate the prospects of using quantum algorithms in agriculture. This opens up new opportunities for solving complex optimization problems in the agricultural sector and improving production efficiency. Further research may be aimed at expanding the application of quantum algorithms in agriculture, including the adaptation of other quantum algorithms, taking into account additional factors in modeling and improving optimization methods. It is also possible to study the application of quantum algorithms in other industries related to resource optimization. In general, the results of this study confirm the potential of quantum algorithms in solving complex optimization problems in agriculture and open up new prospects for the development of the agricultural sector using modern technologies.

References

1. Zahedinejad, E., Zaribafiyan, A.: Combinatorial optimization on gate model quantum computers: A survey. arXiv:1708.05294 (2017)
2. Kandala, A., et al.: Hardware-efficient variational quantum eigensolver for small molecules and quantum magnets. Nature. **549**(7671), 242–246 (2017). https://doi.org/10.1038/nature23879
3. Farhi, E., Neven, H.: Classification with quantum neural networks on near term processors. arXiv:1802.06002 (2018)
4. Sun, K., Sun, X.A.: A two-level distributed algorithm for general constrained non-convex optimization with global convergence. arXiv:1902.07654 (2019)
5. Farhi, E., Goldstone, J., Gutmann, S., Neven, H.: Quantum algorithms for fixed qubit architectures. arXiv:1703.06199 (2017)
6. Nannicini, G.: Performance of hybrid quantum-classical variational heuristics for combinatorial optimization. Phys. Rev. E. **99**(1), 013304 (2019). https://doi.org/10.1103/PhysRevE.99.013304
7. Braine, L., Egger, D. J., Glick, J., Woerner, S.: Quantum algorithms for mixed binary optimization applied to transaction settlement. arXiv:1910.05788 (2019)
8. Vyskocil, T., Pakin, S., Djidjev, H.N.: Embedding inequal-city constraints for quantum annealing optimization. Proc. Int. Workshop Quantum Technol. Optim. Problems, pp. 11–22 (2019). https://doi.org/10.1007/978-3-030-14082-3_2
9. Davis, D., Yin, W.: Convergence rate analysis of several splitting schemes. In: Glowinski, R., Osher, S., Yin, W. (eds.) Splitting Methods in Communication and Imaging, Science and Engineering. Springer, Cham, Switzerland (2017). https://doi.org/10.1007/978-3-319-41589-5_4
10. Ghadimi, E., Teixeira, A., Shames, I., Johansson, M.: Optimal parameter selection for the alternating direction method of multipliers (ADMM): quadratic problems. IEEE Trans. Autom. Control. **60**(3), 644–658 (2015). https://doi.org/10.1109/TAC.2014.2354892
11. Giselsson, P., Boyd, S.: Linear convergence and metric selection for Douglas-Rachford splitting and ADMM. IEEE Trans. Autom. Control. **62**(2), 532–544 (2017). https://doi.org/10.1109/TAC.2016.2564160
12. Wang, Y., Yin, W., Zeng, J.: Global convergence of ADMM in nonconvex nonsmooth optimization. J. Sci. Comput. **78**(1), 29–63 (2019). https://doi.org/10.1007/s10915-018-0757-z
13. Wu, B., Ghanem, B.: P-box ADMM: A versatile framework for integer programming. IEEE Trans. Pattern Anal. Mach. Intell. **41**(7), 1695–1708 (2019). https://doi.org/10.1109/TPAMI.2018.2845842
14. Themelis, A., Patrinos, P.: Douglas-Rachford splitting and ADMM for nonconvex optimization: tight convergence results. SIAM J. Optim. **30**(1), 149–181 (2020). https://doi.org/10.1137/18M1163993

15. Ao F.L.B., Kueng, R., Stilck França, D.: Faster quantum and classical SDP approximations for quadratic binary optimization. arXiv:1909.04613 (2019)
16. Romero, J., Olson, J.P., Aspuru-Guzik, A.: Quantum autoencoders for efficient compression of quantum data. Quantum Sci. Technol. **2**(4), 045001 (2017). https://doi.org/10.1088/2058-9565/aa8072
17. Gilliam, A., Woerner, S., Gonciulea, C.: Grover adaptive search for constrained polynomial binary optimization. arXiv:1912.04088 (2019)

The Detection of a Jump-Like Gaussian Random Disturbance When Appearance Time and Dispersion Are Unknown

Oleg Chernoyarov, Sergey Vybornov, Elena Chernoiarova, and Maria Kholodova

Abstract A single-channel adaptive algorithm for a detection of a fast fluctuating Gaussian random disturbance with the unknown appearance time and dispersion is synthesized. Using the method of local-Markov approximation of the decision determining statistics, or its increment, the closed analytical expressions for the false alarm and missing probabilities are found. By the methods of statistical simulation, it is established that the proposed detector is operable, and the analytical formulas describing the quality of its functioning are in good agreement with the corresponding experimental data in a wide range of values of the analyzed random process parameters.

Keywords Gaussian random disturbance · Unknown appearance time · Unknown dispersion · Maximum likelihood method · Adaptive approach · Detection · Local Markov approximation method · False alarm and signal missing probabilities · Statistical simulation

1 Introduction

In a number of applications of control theory, technical and medical diagnostics, pattern recognition, measurement data processing, etc., it is necessary to solve the problem of statistical analysis of jump-like random disturbances of diagnosed processes and systems [1, 2]. In practice, the diagnosed objects may have stochastic nature or be subjected to external random influences. A jump-like disturbance of

O. Chernoyarov (✉)
National Research University "Moscow Power Engineering Institute", Moscow, Russia

National Research Tomsk State University, Tomsk, Russia
e-mail: chernoyarovov@mpei.ru

S. Vybornov · E. Chernoiarova · M. Kholodova
National Research University "Moscow Power Engineering Institute", Moscow, Russia

A. Gibadullin, S. Gordin (eds.), *Computing Technologies and Applied Mathematics*, Springer Proceedings in Mathematics & Statistics 500,
https://doi.org/10.1007/978-3-031-90546-9_3

such objects is often a random process uniformly occupying the entire operating frequency band and occurring at some a priori unknown moment of time [3–5].

In [6], a technique for detecting a fast-fluctuating bandpass Gaussian random disturbance and measuring its unknown appearance time is considered. It is shown that the use of the proposed approach allows obtaining efficient algorithms for processing random disturbances under the conditions of parametric a priori uncertainty. Below, based on the results obtained in [6], the structure and characteristics of the detector of a Gaussian random disturbance with the unknown appearance time and dispersion are found. It is accomplished under the assumption of the rapidity of Gaussian random disturbance fluctuations and the relative uniformity of its spectral density in a given frequency band. It is demonstrated that the synthesized detector is technically much simpler compared to its analogues and can be implemented on a modern element base. The methods of statistical simulation prove the operability and sufficiently high efficiency of the considered detection algorithm.

2 Problem Statement

Under the jump-like random disturbance, we understand the multiplicative combination like that one:

$$s(t,\lambda_0) = \xi(t)\,\theta(t-\lambda_0), \quad \theta(x) = \begin{cases} 1, & x \geq 0, \\ 0, & x < 0 \end{cases} \tag{1}$$

where λ_0 is the appearance time (i.e., the start) of the disturbance, and $\xi(t)$ is the stationary centered Gaussian random process with the spectral density.

$$G_\xi(\omega) = \frac{\pi D_0}{\Omega}\left[I\left(\frac{\vartheta-\omega}{\Omega}\right) + I\left(\frac{\vartheta+\omega}{\Omega}\right)\right], I(x) = \begin{cases} 1, & |x| \leq 1/2, \\ 0, & |x| > 1/2. \end{cases} \tag{2}$$

In (2), the notations are: ϑ is the central frequency, Ω is the frequency bandwidth, and D_0 is the dispersion (i.e., the average power) of the process $\xi(t)$.

The practically important case under consideration now is when the duration $\tau_0 = T - \lambda_0$ of the disturbance (1) is sufficiently greater than the correlation time $2\pi/\Omega$ of the process $\xi(t)$ (meaning that the fluctuations of the process $\xi(t)$ are "fast"). Thus, the following condition is fulfilled [1, 2, 7]:

$$\tau_0 >> 2\pi/\Omega, (\mu_0 = \tau_0\Omega/2\pi >> 1) \tag{3}$$

We here assume that the registration errors and the noise at the receiver input can be described using Gaussian white noise $n(t)$ with the one-sided spectral density N_0 [3, 8, 9]. As a result, in the presence of the random disturbance (1), one can register the following additive mixture at the observation interval $t \in [0, T]$:

$$x(t) = s(t) + n(t), \tag{4}$$

while, when there is no disturbance, this realization takes the form:

$$x(t) = n(t), \tag{5}$$

Based on the observations (4) and (5), the somehow optimal detection algorithm of the random disturbance (1) should be generated, and its characteristics determined. In this case, it is presupposed that the parameters λ_0 and D_0 are a priori unknown and may take the values from the intervals $[\Lambda_1, \Lambda_2]$ and $[0, \infty)$ at $0 \leq \Lambda_1 < \Lambda_2 < T$.

3 The Detection of the Random Disturbance with the Unknown Appearance Time and Dispersion

Upon detecting a random disturbance (1), we will use an adaptive (the maximum likelihood) approach [10–12]. To this end, using the results from [6], we will write the expression for the decisive statistics (the logarithm of the likelihood ratio functional) when adopting the realization (4) against the alternative (5) as follows

$$L(\lambda, D) = L(\lambda, q) = \frac{q}{N_0(1+q)} \int_{\lambda}^{T} y^2(t)\, dt - \frac{\Omega(T-\lambda)}{2\pi} \ln(1+q). \tag{6}$$

Here λ and D are the current values of the unknown parameters λ_0 and D_0; $q = D/E_N$; $E_N = N_0\Omega/2\pi$ is the average power of the noise $n(t)$ in the bandwidth of the analyzed process $\xi(t)$; and $y(t) = \int_{-\infty}^{\infty} x(t')\, h(t-t')\, dt'$ the output signal of the filter with the transfer function $H(\omega)$ satisfying the condition

$$|H(\omega)|^2 = I[(\omega - \vartheta)/\Omega] + I[(\omega + \vartheta)/\Omega]. \tag{7}$$

Then, the solution in favour of the hypothesis (4) or (5) is made based on the comparison of the magnitude

$$L_m = \max_{\lambda \in [\Lambda_1, \Lambda_2],\, q \geq 0} L(\lambda, q), \tag{8}$$

with the threshold c chosen in accordance with the adopted optimality criterion. If this threshold is surpassed by the magnitude (8), it is assumed that the random disturbance (1) is present in the data under observation; otherwise, it is concluded that it is absent.

In (8), one can perform maximization by q, and it can be done analytically. The resulting decisive rule is then:

$$\hat{L}_m = \max_{\lambda \in [\Lambda_1, \Lambda_2]} \hat{L}(\lambda) \overset{"1"}{\underset{"0"}{\gtrless}} c, \tag{9}$$

where

$$\widehat{L}(\lambda) = \max_q L(\lambda, q) = \frac{1}{N_0} \int_\lambda^T y^2(t)\, dt - \frac{\Omega(T-\lambda)}{2\pi} \left[1 + \ln\left(\frac{1}{E_N(T-\lambda)} \int_\lambda^T y^2(t)\, dt \right) \right]. \tag{10}$$

4 The Characteristics of the Adaptive Detection Algorithm of the Random Disturbance with the Unknown Appearance Time and Dispersion

To quantify the efficiency of the detection, we will use the type 1 (the false alarm) α and type 2 (missing) β error probabilities [10–12].

To determine the probability of a false alarm, we now present the decision determining statistics (10) in the form of

$$\widehat{L}(\lambda) = \widehat{L}(l) = \mu \left[\widehat{L}_1(l) - (1-l) \left(1 + \ln\left(\frac{1}{1-l} \widehat{L}_1(l) \right) \right) \right], \widehat{L}_1(l) = \frac{1}{E_N} \int_l^1 y^2(T\widetilde{t})\, d\widetilde{t}, \tag{11}$$

where

$$l = \lambda/T, \widetilde{t} = t/T, \mu = T\Omega/2\pi \tag{12}$$

are the non-dimensional magnitudes. Then, the false alarm probability should be written as

$$\alpha = P\left[\max_{l \in \left[\widetilde{\Lambda}_1, \widetilde{\Delta}_2\right]} \widehat{L}(l) > c \,|\, x(t) = n(t) \right], \widetilde{\Lambda}_{1,2} = \Lambda_{1,2}/T. \tag{13}$$

Similar to [12], it can be shown that the functional $\widehat{L}_1(l)$ (11) is asymptotically Gaussian at $\mu \to \infty$. Then, when carrying out (3), its complete statistical description is given by the first two moments—that are the mathematical expectation or signal function $\hat{S}_1(l) = \langle \hat{L}_1(l) \rangle$ [13] and the correlation function $\hat{B}_1(l_1, l_2) = \langle \hat{N}_1(l_1) \hat{N}(l_2) \rangle$ of the noise function $\hat{N}_1(l) = \hat{L}_1(l) - \langle \hat{L}_1(l) \rangle$ [13]. Here and further, $\langle \rangle$ stands for the operation of averaging all possible observations of $x(t)$ (if required, under the fixed values of λ_0 and D_0).

Taking into account (3), we can neglect the order of values μ^{-1} and less in (11). Then, when performing the averaging operations, for the characteristics $\widehat{S}_1(l)$ and $\widehat{B}_1(l_1, l_2)$ of the sufficient statistics $\widehat{L}_1(l)$ (11) in the case when $x(t) = n(t)$, we find

$$\widehat{S}_1(l) = 1 - l, \widehat{B}_1(l_1, l_2) = \min(1 - l_1, 1 - l_2)/\mu, l, l_1, l_2 \in \left[\widetilde{\Lambda}_1, \widetilde{\Lambda}_2\right] \quad (14)$$

Let us introduce

$$\varepsilon = 1/\sqrt{\mu}. \quad (15)$$

When performing (3), $\mu >> 1$, so that the magnitude (15) can be considered as a small parameter. Then, using the representations (14), we rewrite the functional $\widehat{L}(l)$ (11) for the case of the absence of a random disturbance (1) as

$$\widehat{L}(l) = \mu\left[1 - l + \varepsilon\widetilde{N}_1(l) - (1 - l)\left(1 + \ln\left(1 + \varepsilon\widetilde{N}_1(l)/(1 - l)\right)\right)\right]. \quad (16)$$

Here $\widetilde{N}_1(l) = \widehat{N}_1(l)\sqrt{\mu}$ is the normalized noise function with the characteristics $\left\langle \widetilde{N}_1(l) \right\rangle = 0$ and $\left\langle \widetilde{N}_1(l_1)\widetilde{N}_1(l_2) \right\rangle = \min(1 - l_1, 1 - l_2)$.

Taking into account the smallness of the magnitude (15), we expand (16) into a Macloren series by the degrees of ε and limit ourselves to the first member of the expansion, depending on the received realization of $x(t)$. Then, in the case when $\varepsilon \to 0$, we get

$$\widehat{L}(l) \approx \widetilde{N}_1^2(l)/2(1 - l), \quad (17)$$

If we make the change of the variables $\widetilde{l} = 1 - l$ and $\widetilde{l} \in \left[1 - \widetilde{\Lambda}_2, 1 - \widetilde{\Lambda}_1\right]$ in (17), then we can obtain.

$$\left\langle \widetilde{N}_1\left(\widetilde{l}\right) \right\rangle = 0, \left\langle \widetilde{N}_1\left(\widetilde{l}_1\right)\widetilde{N}_1\left(\widetilde{l}_2\right) \right\rangle = \min\left(\widetilde{l}_1, \widetilde{l}_2\right) \text{ and } \widehat{L}\left(\widetilde{l}\right) \approx \widetilde{N}_1^2\left(\widetilde{l}\right)/2\widetilde{l}. \quad (18)$$

As the functional $\widehat{L}_1(l)$ (11), as well as the functional $\widehat{N}_1(l)$, is the approximately Gaussian one, it follows from (18) that the process $\widetilde{N}_1\left(\widetilde{l}\right)$ is a standard Wiener random process [14].

The following equation is now introduced.

$$\varphi(u) = \chi(\exp(u)), u \in [U_1, U_2] \quad (19)$$

describing the random process with the probability density $w_\varphi(u) = \exp(-u)/\sqrt{\pi u}$ and the correlation coefficient

$$R_\varphi = \exp(-|u_2 - u_1|), \tag{20}$$

where $\chi(\tilde{l}) = \widetilde{N}_1^2(\tilde{l})/2\tilde{l}$, $u = \ln(\tilde{l})$, $U_1 = \ln(1 - \widetilde{\Lambda}_2)$, and $U_2 = \ln(1 - \widetilde{\Lambda}_1)$. Thus, the false alarm probability (13) can be represented as

$$\alpha = P\left[\sup_{u \in [U_1, U_2]} \varphi(u) > c\right] = 1 - \widehat{P}_N(c), \tag{21}$$

where $\widehat{P}_N(c) = P[\varphi(u) < c]$, $u \in [U_1, U_2]$.

The general expression for the probability $P_\nu(c) = P[\varpi(u) < c]$, $u \in [U_1, U_2]$ of non-crossing the threshold c by the random process $\varpi(u)$ with the probability density $w_\varpi(u) = u^{\nu-1} \exp(-u)/\Gamma(\nu)$ (where $\Gamma(\nu)$ is the gamma-function) and the correlation coefficient $R(\tau)$, which allows the following representation at $\tau \to 0$:

$$R(\tau) = 1 - \delta|\tau| + o(|\tau|), \delta > 0,$$

it was produced in [4]. Using these results, one can now write that

$$P_\nu(c) = \begin{cases} \exp[-\widetilde{m}c^\nu \exp(-c)/\Gamma(\nu)], & c \geq \nu, \\ 0, & c < \nu, \end{cases} \tag{22}$$

where $\widetilde{m} = \delta(U_2 - U_1)$. The accuracy of this approximation increases with $\widetilde{m}$ and c. Considering in (22), as it follows from (19) and (20), that $\nu = 1/2$, $\delta = 1$ and $\widetilde{m} = \ln\left((1 - \widetilde{\Lambda}_1)/(1 - \widetilde{\Lambda}_2)\right)$, we substitute the derivative expression into (21) and get the following for the false alarm probability:

$$\alpha \approx \begin{cases} 1 - \left((1 - \widetilde{\Lambda}_2)/(1 - \widetilde{\Lambda}_1)\right)^{\sqrt{c/\pi}\exp(-c)}, & c \geq 1/2, \\ 1, & c < 1/2. \end{cases} \tag{23}$$

Here, the accuracy of this approximation increases with the relation $(1 - \widetilde{\Lambda}_2)/(1 - \widetilde{\Lambda}_1)$ decreasing and the threshold c increasing, i.e., when the false alarm probability diminishes.

At $c > > 1$, the expression (23) can be simplified so that it gets the form

$$\alpha \approx \sqrt{c/\pi}\exp(-c)\ln\left((1 - \widetilde{\Lambda}_1)/(1 - \widetilde{\Lambda}_2)\right).$$

The probability of missing the appearance time of the random disturbance β is now our concern. By the definition

$$\beta = P\left[\sup \widehat{L}(l) < c \,\middle|\, x(t) = s(t) + n(t)\right], l \in \left[\widetilde{\Lambda}_1, \widetilde{\Lambda}_2\right], \tag{24}$$

where $\widehat{L}(l)$ is determined according to (11).

Following [12], it can be shown that, in the presence of the random disturbance (1) in the observations, the functional $\widehat{L}_1(l)$ included in (11) is asymptomatically Gaussian at $\mu \to \infty$ (12). In this case, when performing (3), for the complete statistical description of this functional, it is sufficient to specify the first two moments—the signal function $\hat{S}_1(l) = \langle \hat{L}_1(l) \rangle$ and the correlation function $\hat{B}_1(l_1, l_2) = \langle \hat{N}_1(l_1)\hat{N}_1(l_2) \rangle$ of the noise function $\hat{N}_1(l) = \hat{L}_1(l) - \langle \hat{L}_1(l) \rangle$, and then $\widehat{L}_1(l) = \widehat{S}_1(l) + \widehat{N}_1(l)$ [13].

We now take (11) and fulfill the averaging by all the possible realizations (4) at the fixed values of the parameters λ_0 and D_0, neglecting the values of the order of smallness μ^{-1} and the lower ones, and get

$$\widehat{S}_1(l) = q_0 \min(1 - l_0, 1 - l) + 1 - l, \tag{25}$$

$$\widehat{B}_1(l_1, l_2) = [q_0(2 + q_0) \min(1 - l_0, 1 - l_1, 1 - l_2) + \min(1 - l_1, 1 - l_2)]/\mu,$$

where $q_0 = D_0/E_N$ and $l_0 = \lambda_0/T$.

While performing (3), we use the definition of the small parameter (15) and the representations (25), and rewrite the decision determining statistics (11) as

$$\widehat{L}(l) = \mu\left\{\widehat{S}_1(l) + \varepsilon \widehat{N}_1(l) - (1 - l)\left[1 + \ln\left(\left(\widehat{S}_1(l) + \varepsilon \widehat{N}_1(l)\right)/(1 - l)\right)\right]\right\}, \tag{26}$$

where $\widehat{N}_1(l) = \widehat{N}_1(l)\sqrt{\mu}$ is the normalized noise function with the characteristics $\left\langle \widehat{N}_1(l) \right\rangle = 0$ and

$$\left\langle \widehat{N}_1(l_1)\widehat{N}(l_2) \right\rangle = q_0(2 + q_0) \min(1 - l_0, 1 - l_1, 1 - l_2) + \min(1 - l_1, 1 - l_2),$$

We now expand (26) into a Macloren series by ε and restrict ourselves to the first member of expansion depending on $x(t)$ (4). At $\varepsilon \to 0$, it allows us to produce.
form:

$$\widehat{L}(l) = \mu\left[\widehat{S}(l) + \varepsilon \widehat{N}(l)\right], \tag{27}$$

where

$$\widehat{S}(l) = \widehat{S}_1(l) - (1-l)\left[1 + \ln\left(\widehat{S}_1(l)/(1-l)\right)\right] \tag{28}$$

is the normalized signal function, and

$$\widehat{N}(l) = \left(\widehat{S}_1(l) - 1 + l\right)\widehat{N}_1(l)/\widehat{S}_1(l) \tag{29}$$

is the normalized noise function of the decision determining statistics $\widehat{L}(l)$.

From (28), one can easily see that the function $\widehat{S}(l)$ reaches its maximum at the point $l = l_0$. It means that the input signal-to-noise ratio (SNR) for the algorithm (9) can be determined as in [12]:

$$\widehat{z}^2 = \mu\widehat{S}^2(l_0)/\left\langle \widehat{N}^2(l_0)\right\rangle = \mu(1-l_0)\left(1 - \ln(1+q_0)/q_0\right)^2. \tag{30}$$

It is assumed that the condition of high a posterior accuracy [11, 12] is achieved during the processing of the realization (4), so that $\widehat{z}^2 >> 1$. In this case, with the probability tending to 1, the position of the greatest maximum $\widehat{l}_m = \mathrm{argsup}\widehat{L}(l)$ of the functional $\widehat{L}(l)$ (27) is within the small δ_q-neighborhood of the point l_0. With the unlimited SNR $\widehat{z}^2$ increasing ($\widehat{z}^2 \to \infty$), the value of the specified neighborhood is

$$\delta_q = \left|\widehat{l}_m - l_0\right| \to 0. \tag{31}$$

Then, when the condition $\widehat{z}^2 >> 1$ is fulfilled, the inequality $\delta_q << 1$ is performed, and, for the normalized signal function (28) and the correlation function of the noise function (29), in neglecting the values of the higher order of smallness than δ_q, we can write the following approximate expressions:

$$\begin{aligned} \widehat{S}(l) \approx\ & (1-l_0)(q_0 - \ln(1+q_0)) + \frac{q_0^2}{1+q_0}\min(0, l_0 - l) \\ & + \left(\frac{q_0}{1+q_0} - \ln(1+q_0)\right)(l_0 - l)\widehat{B}(l_1, l_2) \\ \approx\ & \widehat{\sigma}^2 + \frac{q_0^2}{(1+q_0)^2}[\min(l_0 - l_1, l_0 - l_2) + q_0(2+q_0)\min(0, l_0 - l_1, l_0 - l_2)] \\ & + \frac{q_0^2}{1+q_0}[\min(0, l_0 - l_1) + \min(0, l_0 - l_2) + l_1 + l_2 - 2l_0], \end{aligned} \tag{32}$$

where $\widehat{\sigma}^2 = q_0^2(1-l_0)$.

Now it is time to pass from the decision determining statistics (10), (27) to the decision determining statistics expressed as

$$\widehat{\upsilon}(l) = \left[\widehat{L}(l) - \widehat{L}_0\right] / \widehat{\sigma}\sqrt{\mu}, \tag{33}$$

where $\widehat{L}_0 = \widehat{L}(l_0)$ is the asymptotically Gaussian random magnitude (with μ (12) increasing), and it has the characteristics $\langle \hat{L}_0 \rangle = \mu(1 - l_0)(q_0 - \ln(1 + q_0))$ and $\left\langle (\hat{L}_0 - \langle \hat{L}_0 \rangle)^2 \right\rangle = \mu \hat{\sigma}^2$, while $\widehat{\sigma}^2$ is determined in the same way as in (32).

Using the representation (33), we now rewrites the missing probability (24) of the appearance time of the random disturbance (1) in the form

$$\beta = P\left[\sup_{l \in \left[\widetilde{\Lambda}_1, \widetilde{\Lambda}_2\right]} \widehat{\upsilon}(l) < \left(c - \widehat{L}_0\right) / \widehat{\sigma}\sqrt{\mu} \,|\, x(t) = s(t) + n(t)\right]. \tag{34}$$

When performing (31) of (32) for the mathematical expectation and the correlation function of the function $\widehat{\upsilon}(l)$ (33), we get

$$\langle \widehat{\upsilon}(l) \rangle \approx \widehat{S}_\upsilon(l), \ \langle [\widehat{\upsilon}(l_1) - \langle \widehat{\upsilon}(l_1) \rangle] [\widehat{\upsilon}(l_2) - \langle \widehat{\upsilon}(l_2) \rangle] \rangle \approx \widehat{B}_\upsilon(l_1, l_2), \tag{35}$$

where

$$\widehat{S}_\upsilon(l) = \frac{\sqrt{\mu}}{\widehat{\sigma}} \begin{cases} \widehat{A}_1(l - l_0), & l < l_0, \\ -\widehat{B}_1(l - l_0), & l \geq l_0, \end{cases}$$

$$\widehat{B}_\upsilon(l_1, l_2) = \frac{1}{\widehat{\sigma}^2} \begin{cases} \widehat{A}_2 \min(|l_1 - l_0|, |l_2 - l_0|), & (l_1 - l_0)(l_2 - l_0) \geq 0, \\ 0, & (l_1 - l_0)(l_2 - l_0) < 0, \end{cases}$$

$$\widehat{A}_1 = \ln(1 + q_0) - \frac{q_0}{1 + q_0}, \widehat{B}_1 = q_0 - \ln(1 + q_0), \widehat{A}_2 = \begin{cases} q_0^2 / (1 + q_0)^2, & l_1 < l_0, l_2 < l_0, \\ q_0^2, & l_1 \geq l_0, l_2 \geq l_0. \end{cases} \tag{36}$$

From (36) it is not difficult to see that the values of the process $\widehat{\upsilon}(l)$ at the intervals $[l_0 - \delta_q, l_0)$ and $[l_0, l_0 + \delta_q]$ are non-correlated. Therefore, due to the asymptomatic Gaussian nature of this process, its values are approximately statistically independent (at the final values of μ). Then, similar to [13], the probability $\widehat{F}_\upsilon(x) = P\left[\sup_{l \in [l_0 - \delta_q, l_0 + \delta_q]} \widehat{\upsilon}(l) < x\right]$ can be represented as

$$\widehat{F}_\upsilon(x) = \widehat{F}_1(x)\,\widehat{F}_2(x). \tag{37}$$

where

$$\widehat{F}_1(x) = P\left[\sup_{l \in [l_0 - \delta_q,\, l_0)} \widehat{\upsilon}(l) < x\right], \widehat{F}_2(x) = P\left[\sup_{l \in [l_0,\, l_0 + \delta_q]} \widehat{\upsilon}(l) < x\right]. \tag{38}$$

Using (37), for the required probability (34), one gets

$$\beta = \beta(c) = \int_{-\infty}^{c/\widehat{\sigma}\sqrt{\mu}} \widehat{F}_\upsilon\left(c/\widehat{\sigma}\sqrt{\mu} - x\right) \widehat{w}_0(x)\, \mathrm{d}x =$$
$$= \int_{-\infty}^{c/\widehat{\sigma}\sqrt{\mu}} \widehat{F}_1\left(c/\widehat{\sigma}\sqrt{\mu} - x\right) \widehat{F}_2\left(c/\widehat{\sigma}\sqrt{\mu} - x\right) \widehat{w}_0(x)\, \mathrm{d}x\ , \tag{39}$$

where

$$\widehat{w}_0(x) = \exp\left[-(x - \widehat{z})^2/2\right]/\sqrt{2\pi} \tag{40}$$

is the probability density of the Gaussian random value $\widehat{L}_0/\widehat{\sigma}\sqrt{\mu}$, and $\widehat{z}$ is determined based on (30).

Our next step is to introduce

$$\widehat{\upsilon}_1(l) = x - \widehat{\upsilon}(l_0 - l),\ \widehat{\upsilon}_2(l) = x - \widehat{\upsilon}(l - l_0), l \in \left[l_0, l_0 + \delta_q\right]. \tag{41}$$

It can be shown, as in [15], that in the subregion $l \geq l_0$ for the processes $\widehat{\upsilon}_i(l)$, $i = 1$, 2 (41), the conditions of the Doob theorem are fulfilled [16]. According to this theorem, the processes $\widehat{\upsilon}_i(l)$, $i = 1$, 2 are Gaussian Markov random processes. At that, their drift coefficient $\widehat{a}_i$ and diffusion coefficient $\widehat{b}_i$, as can be inferred from (35) and (36), are determined as follows:

$$\widehat{a}_1 = \sqrt{\frac{\mu}{1 - l_0}}\left(\frac{\ln(1 + q_0)}{q_0} - \frac{1}{1 + q_0}\right),\ \widehat{a}_2 = \sqrt{\frac{\mu}{1 - l_0}}\left(1 - \frac{\ln(1 + q_0)}{q_0}\right),$$
$$\widehat{b}_1 = \frac{1}{(1 + q_0)^2(1 - l_0)},\ \widehat{b}_2 = \frac{1}{1 - l_0} \tag{42}$$

Using the ratio (42), the required distribution functions (37) can be written as, similar to [13]

$$F_i(x) = P\left[\sup_{l \in [l_0, l_0+\delta]} \quad v_i(l) > 0\right] = \int_0^{\infty} w_i(y, l_0 + \delta)\, dy, \tag{43}$$

where $w_i(y, l)$ are the probability densities of that the processes $\widehat{v}_i(l)$, $i = 1, 2$ taking the values $v_i(l_0) = x$ at the starting moment $l = l_0$, reach the values $v_i(l_0 + \delta_q) = y$ at the end of the analysis interval $l = l_0 + \delta_q$, while $0 < \underset{l \in [l_0, l_0+\delta_q]}{v_i(l)} < \infty$.

The analytical expressions for the probability densities $w_i(y, l)$ and $w_{2i}(y, \widetilde{\eta})$ in (43) are the solutions of the Fokker-Planck-Kolmogorov equations [14] with the drift coefficients $\widehat{a}_i$ and diffusion coefficient $\widehat{b}_i$ (42) under the initial conditions $w_i(y, l_0) = \delta(x - y)$ and boundary conditions $w_i(0, l) = w_i(\infty, l) = 0$. We solve these equations by the method of a characteristic function and take into account the specified boundary conditions using the reflection method with a change of signs [14], as described in [13]. And then we integrate the found solutions and replace the results in (38), for the functions (37), getting as a result:

$$\widehat{F}_v(x) = f\left(x, \widehat{a}_1, \widehat{b}_1, \delta_q\right) f\left(x, \widehat{a}_2, \widehat{b}_2, \delta_q\right). \tag{44}$$

if $x \geq 0$, and $F_v(x) = 0$ if $x < 0$. Here

$$f(x, a, b, \delta) = \Phi\left(\frac{a\delta + x}{\sqrt{b\delta}}\right) - \exp\left(-\frac{2ax}{b}\right) \Phi\left(\frac{a\delta - x}{\sqrt{b\delta}}\right). \tag{45}$$

In practice, it is rather difficult to calculate the probability (24) using the formulas (39) (40), (44), (45) because there is no sound methodology for determining the value of δ_q. However, in the conditions of high a posterior accuracy, when $\delta_q < < 1$ and $\widetilde{z}^2 > > 1$ (30), it is possible to use the asymptomatic (at $\delta_q \to 0$) approximation of the function like $f(x, a, b, \delta) = 1 - \exp(-2ax/b)$ in (44) instead of (45). Then, the expression (44) is simplified and takes the form

$$\widehat{F}_v(x) = \left[1 - \exp\left(-2\widehat{a}_1 x/\widehat{b}_1\right)\right]\left[1 - \exp\left(-2\widehat{a}_2 x/\widehat{b}_2\right)\right]. \tag{46}$$

By substituting (40) and (46) into (39) and integrating the received expression by the variable x, we finally get:

$$\beta=\Phi\left(\frac{c}{\widehat{\sigma}\sqrt{\mu}}-\widehat{z}\right)-\exp\left[\frac{2\widehat{a}_1}{\widehat{b}_1}\left(\frac{\widehat{a}_1}{\widehat{b}_1}+\widehat{z}-\frac{c}{\widehat{\sigma}\sqrt{\mu}}\right)\right]\Phi\left(\frac{c}{\widehat{\sigma}\sqrt{\mu}}-\widehat{z}-\frac{2\widehat{a}_1}{\widehat{b}_1}\right)$$
$$-\exp\left[\frac{2\widehat{a}_2}{\widehat{b}_2}\left(\frac{\widehat{a}_2}{\widehat{b}_2}+\widehat{z}-\frac{c}{\widehat{\sigma}\sqrt{\mu}}\right)\right]\Phi\left(\frac{c}{\widehat{\sigma}\sqrt{\mu}}-\widehat{z}-\frac{2\widehat{a}_2}{\widehat{b}_2}\right)+ \tag{47}$$
$$+\exp\left[2\left(\frac{\widehat{a}_1}{\widehat{b}_1}+\frac{\widehat{a}_2}{\widehat{b}_2}\right)\left(\frac{\widehat{a}_1}{\widehat{b}_1}+\frac{\widehat{a}_2}{\widehat{b}_2}+\widehat{z}-\frac{c}{\widehat{\sigma}\sqrt{\mu}}\right)\right]\Phi\left(\frac{c}{\widehat{\sigma}\sqrt{\mu}}-\widehat{z}-\frac{2\widehat{a}_1}{\widehat{b}_1}-\frac{2\widehat{a}_2}{\widehat{b}_2}\right),$$

where $\Phi(x)=\int_{-\infty}^{x}\exp(-u^2/2)\,du/\sqrt{2\pi}$ is the probability integral. The accuracy of the formula (47) increases with μ (12), $\widehat{z}$ (30) and c.

5 The Results of the Statistical Simulation

A computer statistical simulation of the performance of the detector (9) has been performed to determine the functionality of the algorithm (9), as well as the applicability limits of the asymptomatic formulas (23), (47) for its characteristics.

In the process of simulation, at the interval $\widetilde{t}\in\left[\widetilde{\Lambda}_1,1\right]$ (12), in the discrete moments of time $\widetilde{t}_j=\widetilde{\Lambda}_1+j\Delta\widetilde{t}$, $j=\overline{0,\mathrm{int}\left(\left(1-\widetilde{\Lambda}_1\right)/\Delta\widetilde{t}\right)}$, for the given values of μ, q, l_0, the non-dimensional samples $\widetilde{y}_j=y\left(T\widetilde{t}_j\right)\sqrt{T/N_0}$ (6) have been generated, as described in [13]. Here int(·) is the integer part. According to the generated samples $\widetilde{y}_{jk}$, and following [13], the samples $\widehat{L}_{1n}=\widehat{L}_1(n\Delta l)$ of the random process $\widehat{L}_1(l)$ (11) are calculated, and then the ones for the decision determining statistics $\widehat{L}_n=\widehat{L}(n\Delta l)$ in the region $\left[\widetilde{\Lambda}_1,\widetilde{\Lambda}_2\right]$ where Δl is the discrete step of change of the current value of the normalized appearance moment (12). The discretionary step by the variable $\widetilde{t}$ (12) is then chosen to be $\Delta\widetilde{t}=0.01/\mu$, and, by the variable l, it is $\Delta l=10^{-3}$. As a result, the relative root-mean-square error of the stepped approximation of the functional $\widehat{L}(l)$ (11), that is based on the generated samples, can be calculated using the method [5], and it does not exceed 10%.

In each test performed (for each generated realizations of $\widehat{L}(l)$) the presence or absence of a random disturbance (1) in the observations is determined by comparing the maximum sample $L_{\max}=\max\limits_n L_n$ with the chosen threshold c: $L_{\max}\ \substack{>\\<}\ c$. Furthermore, the experimental detection and measurement characteristics as described in [13] are determined by direct averaging by all processed realizations $\widehat{L}(l)$.

Figures 1 and 2 present some of the results obtained in the statistical simulation process and the corresponding theoretical dependencies. In order to obtain each experimental value, at least $5\cdot10^4$ realizations of $\widehat{L}(l)$ (11) have been processed at the fixed $\widetilde{\Lambda}_1=0.25$, $\widetilde{\Lambda}_2=0.75$ and $l_0=\left(\widetilde{\Lambda}_1+\widetilde{\Lambda}_2\right)/2$. Thus, the deviation of the

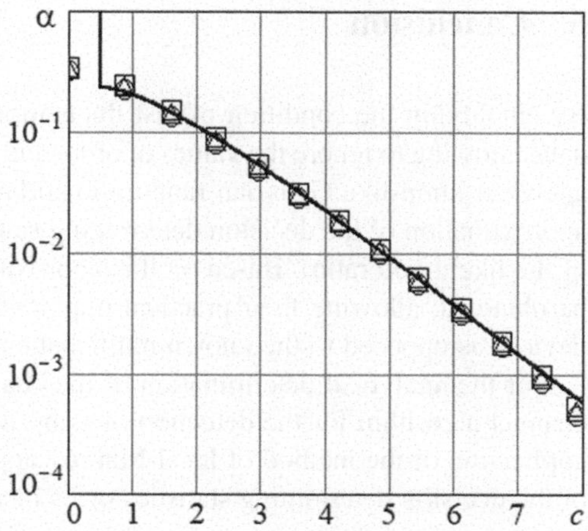

Fig. 1 The dependencies of the false alarm probability when the jump-like disturbance is detected

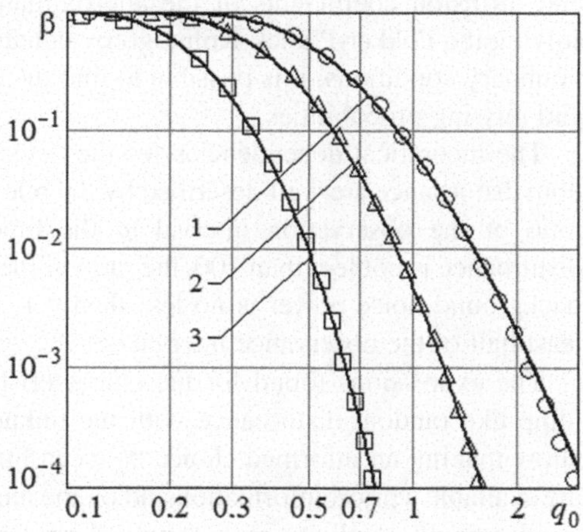

Fig. 2 The dependencies of the missing probability when the jump-like disturbance is detected

confidence interval boundaries from the experimental data appears to be no greater than 10% with the probability of 0.9.

In Fig. 1, the continuous line shows the theoretical dependency $\alpha = \alpha(c)$ (23) of the false alarm probability (13). In Fig. 2, the continuous lines demonstrate the theoretic dependencies $\beta = \beta(q_0)$ (47) of the missing probability (24) at $\mu = 100$ (the curve 1), $\mu = 200$ (the curve 2) and $\mu = 500$ (the curve 3).

The experimental values of the detection characteristics for μ equal to 100, 200 and 500 are indicated in Figs. 1 and 2 by circles, triangles and squares, respectively. The threshold value c when determining the probability β (47) has been selected in accordance with the given false alarm level $\alpha = 0.01$ and following the formula (23).

As it follows from the Figs. 1 and 2, the obtained theoretical dependencies (23) and (47) for the probabilities α (13) and β (24) satisfactorily approximate the experimental data, at least when $\mu \geq 100$ ($\mu_0 \geq 50$) and $q_0 \geq 0.1$, $\widetilde{\Lambda}_1 \geq 0.25$, $\widetilde{\Lambda}_2 \leq 0.75$.

6 Conclusion

By employing the condition of fast fluctuations in the realizations of the observed data, allowing to ignore the values of order and less than the time of correlation of the analyzed jump-like Gaussian random disturbance, one can find a relatively simple approximation of the decision determining statistics (the logarithm of the functional of the likelihood ratio). Based on this approximation, new detection algorithms can be obtained, allowing their practical implementation in the form of single-channel devices as opposed to the known multichannel equipment.

For the analytical determination of the characteristics of the synthesized single-channel algorithm for the detection of jump-like Gaussian random disturbance, the application of the method of local-Markov approximation, using the approximation of the decision determining statistics or its increment through the Gaussian Markov processes of a diffusion type, proves to be effective. Then, by determining the drift and diffusion coefficients of the approximating Gaussian Markov processes and solving the Fokker-Planck-Kolmogorov equation with the corresponding initial and boundary conditions, it is possible to find the closed expressions for the false alarm and missing probabilities.

The theoretical dependencies for the detection characteristics of jump-like random disturbance are well described by the relevant experimental data. It is true if the ratio of the observation interval to the time of the correlation of the random disturbance is not less than 100, the ratio of the disturbance dispersion to the average background noise power is no less than 0.1, and the disturbance itself takes up at least half of the observance interval.

The expressions found for the characteristics of the detection algorithm of the jump-like random disturbance with the unknown appearance time and dispersion allow making an informed choice between this and other algorithms depending on the available a priori information and on the requirements imposed on the quality and the degree of simplicity of its technical implementation.

The algorithm considered for the detection of a random disturbance can be practically implemented using a modern element base in real time scale in both analogue and digital versions.

Acknowledgements This research was financially supported by the Russian Science Foundation (research project No. 24-11-00191).

References

1. Basseville, M., Benveniste, A. (eds.): Detection of Abrupt Changes in Signals and Dynamical Systems. Springer, Berlin (1986)
2. Golpaiegani, L.A., Kalashnikov, K.S., Salnikova, A.V.: Stochastic Signals with Unknown Parameters. Detection and Measurement of the Abrupt Change of the Fast-Fluctuating Random

Processes Under Conditions of A Priori Uncertainty. Lap Lambert Academic Publishing, Moscow (2019)
3. Klovsky, D.D.: Transmitting discrete message over a radio channels. Radio i Svyaz', Moscow (1982)
4. Bakut, P.A. (ed.): Signal Detection Theory. Radio i Svyaz', Moscow (1984)
5. Zakharov, A.V., Trifonov, A.P., Pronyaev, E.V.: Detection of step random disturbance. J. Comput. Syst. Sci. Int. **40**(6), 869–877 (2001)
6. Chernoyarov, O., Vybornov, S., Chernoiarova, E., Kholodova, M.: Detecting and estimating the moment of appearance of an abrupt Gaussian random disturbance under fast fluctuations of the realization of observed data. Lecture Notes in Networks and Systems, 2025. (in press)
7. Hsu, S.T., Whitter, R.J., Mead, S.A.: Physical model for burst noise in semiconductor devices. Sold-State Electron. **13**(7), 1055–1071 (1970)
8. Kadota, T.T., Zakai, M., Ziv, J.: Mutual information on the white Gaussian channel with and without feedback. IEEE Trans. Inf. Theory. **17**(7), 368–371 (1971)
9. Baranov, V.N., Bel'sky, M.E.: The algorithms for estimating the moment of abrupt change of a Gaussian sequence and determining the variance of the resulting estimate. J. Comput. Syst. Sci. Int. **4**, 19–32 (1999)
10. Helstrom, C.W.: Statistical Theory of Signal Detection. Pergamon Press, Oxford (2013)
11. Van Trees, H.L., Bell, K.L., Tian, Z.: Detection, Estimation, and Modulation Theory, Part I. Detection, Estimation, and Filtering Theory, 2nd edn. Wiley, New York, NY (2013)
12. Trifonov, A.P., Nechaev, E.P., Parfenov, V.I.: Detection of Stochastic Signals with Unknown Parameters. Voronezh State University, Voronezh (1991)
13. Chernoyarov, O.V., Min, S.S.T., Salnikova, A.V., Shakhtarin, B.I., Artemenko, A.A.: Application of the local Markov approximation method for the analysis of information processes processing algorithms with unknown discontinuous parameters. Appl. Math. Sci. **8**(90), 4469–4496 (2014)
14. Dynkin, E.B.: Theory of Markov Processes. Dover Publications Inc., New York, NY (2006)
15. Kailath, T.: Some integral equations with nonrational kernals. IEEE Trans. Inf. Theory. **12**(4), 442–447 (1966)
16. Doob, J.L.: Stochastic Processes. Wiley, New York, NY (1953)

Mathematical Analysis of Disturbing Forces in the Air Gap of an Asynchronous Motor Using Finite Element Modeling

Artem Ermolaev, Aleksandr Plekhov, and Dmirtiy Titov

Abstract The article presents a mathematical theory of the synthesis of a finite element model for the study of electromagnetic and magnetomechanical processes in electrical machines. A complete drawing of the electric machine section in a plane perpendicular to the rotation axis was performed when combining the rotor and stator elements into separate geometric objects. A complete drawing of the electric machine section allows for research with asymmetric geometry of its elements. The applicability of the system of equations solved at each boundary to which a mechanical load is applied is proved. The parameters of multiphysical finite element modeling are determined, which perform numerical calculations: magnetic forces driving the rotor, radial forces causing deformation of the stator, dynamic parameters of the electric machine elements, oscillatory processes. Examples of setting up multiphysical solvers of the modeling environment are given. Mechanical loads acting from magnetic forces on the stator teeth are revealed.

Keywords Electric machine · Electromagnetic processes · Finite element model · Deformation · Vibrations

1 Introduction

The operation of electric machines in electromechanical systems, such as electric generators with variable electric load and rotation frequency, is caused by magnetic fields. The latter create radial magnetic forces and stator deformations, which are mechanical waves of a certain frequency, amplitude and spatial order [1–3]. Magnetic vibration can be reduced by effectively controlling the parameters of the motor supply voltage or the current of the generator stator windings, which requires knowledge of the relationship between these physical quantities [4, 5]. Interrelated mechanical, acoustic and electromagnetic calculations of processes occurring in a

A. Ermolaev · A. Plekhov (✉) · D. Titov
Nizhny Novgorod State Technical University n.a. R.E. Alekseev, Nizhny Novgorod, Russia

A. Gibadullin, S. Gordin (eds.), *Computing Technologies and Applied Mathematics*, Springer Proceedings in Mathematics & Statistics 500,
https://doi.org/10.1007/978-3-031-90546-9_4

rotating electric machine are advisable to carry out in finite element multiphysical design environments.

Electromechanical energy conversion systems at variable rotation frequency of the primary combustion engine or wind engine contain a DC link, while adjustable electric drives are based on the use of controlled rectifiers [6, 7]. Simulation dynamic models with lumped parameters are used to study the processes in these systems. These models allow forming the disturbance functions of the specified voltages, currents and moments of electric machines when their electrical and/or mechanical loads change. However, the study of the processes of forming vibration-disturbing forces and moments on the structural elements of electric machines is carried out by means of finite element models [8, 9]. The use of such models, given the multiphysics of their software platform, allows studying the spectral compositions and modes of elastic deformation displacements of the elements of the magnetic circuit of an electric machine when the load on its shaft, amplitude, harmonic composition and asymmetry parameters of the stator currents change.

2 Materials and Methods

The object of the study is a model of an asynchronous electric motor with a squirrel-cage rotor, which was developed in the multiphysical finite element design environment "COMSOL Multiphysics."

The numerical model is implemented using the example of an asynchronous electric motor AIR180M4U3 with a power of 30 kW and a synchronous speed of 1500 rpm. The main geometric and technical parameters of the motor are: effective voltage, rated current, rated torque, rated power, synchronous speed, number of pole pairs, outer stator core diameter, inner stator core diameter, stator and rotor core length, air gap width, stator back height, stator slot height, rotor slot height, stator slot width, rotor slot width, number of stator slots, number of rotor slots. Each element of the model is assigned a material with specified physical properties.

In the process of creating the model, a complete drawing of the electric machine cross-section was made, which allows us to study the influence of supply voltage asymmetry on changes in mechanical stresses and deformations of its individual elements. Drawing provides the ability to take into account the geometry of the machine and build a calculation grid.

In addition, at the stage of creating the geometry of the numerical model of the engine, the elements related to the rotor were combined into a separate geometric object (domain) "Rotor" and, similarly, the elements related to the stator were combined into an object (domain) "Stator." This stage is necessary to create an identical pair—a boundary on which the continuity conditions are taken into account, which is necessary when solving problems with a moving grid.

The rotational motion of the rotor is determined by the system of Eq. (10):

$$\begin{cases} \dfrac{d\omega}{dt} = \dfrac{T - T_l}{J_r} \\ \omega = \dfrac{d\varphi}{dt} \end{cases}, \quad (1)$$

where T is the electromagnetic torque of the engine, T_l is the load torque, φ is the rotor rotation angle, ω is the rotor angular velocity, J_r is the rotor moment of inertia.

The rotor moment of inertia was determined numerically using a special module called "Mass Properties." This function is especially useful when investigating motors with dynamic rotor eccentricity or with uneven load.

The main module that performs electromagnetic and magnetomechanical calculations using Maxwell's equations is the Rotating Machinery. The module solves the following system of equations:

$$\begin{cases} \nabla \times H = J \\ B = \nabla \times A \\ \nabla \cdot D = \rho \\ E = -\dfrac{\partial A}{\partial t} \\ \nabla \cdot A = 0 \end{cases}, \quad (2)$$

where ∇ is the nabla operator, which determines the partial derivatives of a physical quantity with respect to coordinates, H is the magnetic field strength, J is the electric current density, B is the magnetic field induction, A is the magnetic vector potential, D is the electric field induction, E is the electric field strength, ρ is the volume density of the external electric charge, t is time.

The first equation of this system is an expression of the magnetic field circulation theorem. The second equation is auxiliary and establishes a correspondence between the physical meanings of the vector magnetic potential and the magnetic flux density (magnetic induction). The third expression states that the electric charge is the source of electric induction (Gauss's law). The fourth equation is an expression of Faraday's law of induction, which states that a change in magnetic induction generates a vortex electric field. Finally, the fifth expression is Gauss's law for the magnetic field, which denies the existence of magnetic charges. Ampere's law for air domains, structural steel and copper has the following boundary conditions:

$$\begin{cases} J = \sigma E \\ B = \mu_0 \mu H \\ D = \varepsilon_0 \varepsilon E \end{cases}. \quad (3)$$

Thus, the given boundary conditions assume that there is no magnetic saturation in the specified domains, and magnetization is carried out according to a linear law. The values of the coefficients σ, μ, ε in the calculations are taken from the tables of material properties.

The calculation of the electromagnetic torque is carried out by the Arkkio method using the Arkkio Torque Calculation block. This method allows calculating the torque by volume integration of the product of the components of magnetic induction in the air gap of the machine. To calculate the torque T, the domains of the air gap between the stator and the rotor of the electric motor were identified. The expression for calculating the torque is as follows:

$$T = l \cdot n_{\text{sec}} \int_{\Omega} \frac{F_{\phi} r_A}{r_{out} - r_{in}} dV; \; F_{\phi} = \frac{B_r B_{\phi}}{\mu_0}, \tag{4}$$

where l is the length of the stator and rotor core, n_{sec} is the number of symmetric sectors, Ω is the auxiliary Arcchio function limiting the calculation region between r_{in} and r_{out}, F_{ϕ} is the azimuthal force density, r_{A} is the specified air gap radius, r_{in} is the inner radius of the air gap, r_{out} is the outer radius of the air gap, V is the tangential velocity on the rotor surface, B_r is the radial component of magnetic induction, B_{ϕ} is the azimuthal (rotating) component of magnetic induction.

The calculation of magnetic forces driving the rotor is carried out using the "Force Calculation" block. Therefore, for calculations within this block, the rotor domains were selected: its core and shaft. The expressions for calculating magnetic forces and moments are as follows:

$$F = \int_{\partial\Omega} dnTdS; \; \tau = \int_{\partial\Omega} d(r - r_0) \times (nT)dS; \; \tau_{ax} = \frac{r_{ax}}{|r_{ax}|} \cdot \tau, \tag{5}$$

where T is the stress tensor, τ_{ax} is the torque relative to the axis of rotation, r_{ax} is the position of the axis of rotation.

In general, the electromagnetic torque according to the Arkkio method is calculated as follows

$$T = l \cdot \int_0^{2\pi} F_{\phi} r_A dV, \tag{6}$$

while the average torque is determined by the formula:

$$T_{aver} = \frac{l}{r_{out} - r_{in}} \int F_{\phi} r_A dS, \tag{7}$$

where $dS = rdrd\varphi$ is the differential area of the air gap in the polar coordinate system.

In addition, the direction and position of the rotation axis and the point relative to which the torque calculation was performed were specified in this block, as shown in the figure. Here, the value of e_r corresponds to the radius of the static eccentricity—the value by which the rotor axis is shifted relative to the stator axis (by default, equal to 0).

The stator winding of an AC machine, when solving problems in a 2D formulation, can be described by the "Coil" block, designed to describe electromagnetic coils. Currents or voltages of windings can be specified as constant values, as functions, and by approximating a set of points. In the developed numerical model, stator windings are powered by AC voltage sources. The use of voltages as input signals is necessary for the correct description of transient processes of the engine, otherwise there will be no increase in the current amplitude during startup and change in load torque. The time functions of phase voltages are specified analytically.

The equation for calculating currents in phase windings using the "Coil" blocks will look like:

$$J_e = \frac{NI_{cir}}{A} e_{coil}, \tag{8}$$

where e_{coil} is the vector of directions of current flow in the coil, J_e is the density of the current generated by an external source, I_{cir} is the phase current.

A short-circuited rotor winding of an asynchronous electric motor can also be described by several "Coil" blocks, but the voltage must be selected as the source of their excitation, otherwise the current will not flow through the windings. The initial voltage is selected equal to 0 V. The equation for calculating the current in the rotor winding will be as follows:

$$J_e = \frac{N(V_{coil} + V_{ind})}{AR_{coil}} e_{coil}, \tag{9}$$

where R_{coil} is the active resistance of the coil, V_{coil} is the voltage applied to the coil, V_{ind} is the induced voltage calculated by integrating the electric field along the coil.

Solvers must be configured correctly to obtain an accurate solution to the problem for the numerical model. Although the main problem is solved in the "Time Dependent" time domain, the problem must first be solved at t = 0. In this case, the problem is solved in statics in two main stages. In the first stage, the "Coil Geometry Analysis" problem is solved for all coils and conductors, i.e. the directions of all currents are determined. In this case, the problem is solved only for the "Rotating Machinery, Magnetic" module. At the second stage, the problem that determines the vector potential field for future research is solved directly—"Stationary."

In the Solver Configurations menu, the settings remain mostly unchanged. In the Stationary Solver menu, you can set the permissible solution error (the smaller it is, the more accurate the result will be). In the Direct menu, you should select the MUMPS or PARDISO solver, and the Nonlinear Solver "Fully Coupled" is necessary because the model contains materials with nonlinear magnetic properties. For this solver, it is advisable to increase the number of iterations to improve convergence, and use the Automatic (Newton) or Constant (Newton) method.

After receiving the results of the Stationary study, you can begin the study in the time domain. First, you need to set the time range and time step with which the results will be displayed. The plug-ins for this study are selected similarly to the previous step.

After the electromagnetic fields, torque and Lorentz forces have been investigated and the corresponding solution of the numerical model in the time domain has been obtained, it is possible to proceed to the investigation of mechanical forces and deformations caused by these forces. The settings for this task (and in fact a whole series of tasks) are implemented using the "Solid Mechanics" module. This module is designed to investigate mechanical forces, stresses, deformations, oscillation processes, as well as fluidity and fracture processes in solids. To work with this module, only domains related to the stator core are selected. In this case, the following system of equations is solved:

$$\begin{cases} 0 = \nabla\cdot(FS)^{\mathrm{T}} - F_V \\ F = I + \nabla u \end{cases}, \tag{10}$$

where F is the deformation gradient, S is the second Piola-Kirchhoff stress tensor, F_V is the volume force vector, I is the identity matrix, u is the displacement, and ∇ is the velocity of structural displacements.

The main boundary conditions imposed on the material are the conditions of its linear dependence of rigidity on deformation. In this case, the formulation of the conditions itself will have the following form:

$$\begin{aligned} &S = S_{inel} + S_{el}, \\ &\varepsilon_{el} = 0,5\cdot\left(F_{el}^{T}F_{el} - I\right), \\ &F_{el} = FF_{inel}^{-1} \\ &S_{el} = J_i F_{inel}^{-1}(C : \varepsilon_{el})F_{inel}^{-T}, \\ &S_{inel} = S_0 + S_{ext} + S_q, \\ &\varepsilon = 0,5\left[(\nabla u)^T + \nabla u + (\nabla u)^T\nabla u\right], \\ &C = C(E, \nu), \end{aligned} \tag{11}$$

where S_{inel} is the second Piola-Kirchhoff stress tensor for inelastic strains, S_{el} is the second Piola-Kirchhoff stress tensor for elastic strains, ε_{el} is the Green-Lagrange elastic strain, F_{el} is the elastic strain gradient, F_{inel} is the inelastic strain gradient, $J_i = det(F_{inel})$ is the volume coefficient, C is the right Cauchy-Green strain tensor, S_0 is the initial stress tensor, S_{ext} is the external stress tensor, S_q is the stress tensor associated with losses in hyperelastic materials, ε is the Cauchy-Green strain, E is Young's modulus, ν is Poisson's ratio.

The mechanical loads acting on the magnetic forces were specified using a series of "Boundary Load" blocks, i.e. distributed loads acting on the stator teeth. It is known that the distributed force f acting on the stator core is determined by the expression:

$$f = \frac{B^2}{2\mu_0}. \quad (11)$$

In order to correctly calculate the force f, within the framework of the study "Rotating Machinery. Magnetic" a probe "Boundary Probe" was assigned to each outer boundary of the stator teeth, recording the average integral magnetic field at each moment of time. The obtained diagrams $f = B(t)$ for each tooth is recorded in the form of tabular data, and then, based on their approximation, signals of radial magnetic forces acting on each tooth are formed. At each boundary to which a mechanical load is applied, the following system of equations is solved:

At each boundary to which a mechanical load is applied, the following system of equations is solved:

$$\begin{cases} S \cdot n = F_A \\ F_A = -pn \end{cases}, \quad (12)$$

where F_A is the deformation gradient, p is the pressure on the tooth (radial magnetic force).

The last set of conditions necessary to obtain a correct solution of the numerical model is the imposition of boundary conditions of sliding on at least 2 mutually perpendicular faces of the stator core. The fact is that when using the default boundary conditions, the stator core is a body not fixed to anything, and asymmetric radial magnetic forces will drag it far beyond the working space. The use of the condition of rigid fixation for the stator will affect both its rigidity and the forms of its oscillations.

The first study to be conducted for the Solid Mechanics module will be aimed at assessing deformations and mechanical stresses in the time domain. The time range and time step are configured similarly to the corresponding study of electromagnetic fields and forces.

The second study using the Solid Mechanics module will be aimed at determining the oscillation modes and their corresponding frequencies. The oscillation mode is a curve showing the position of the points of the oscillatory system relative to the equilibrium position at a fixed point in time [11, 12]. Thus, for the stator core, the oscillations will have the shape of a regular r-gon, as shown in the Fig. 1.

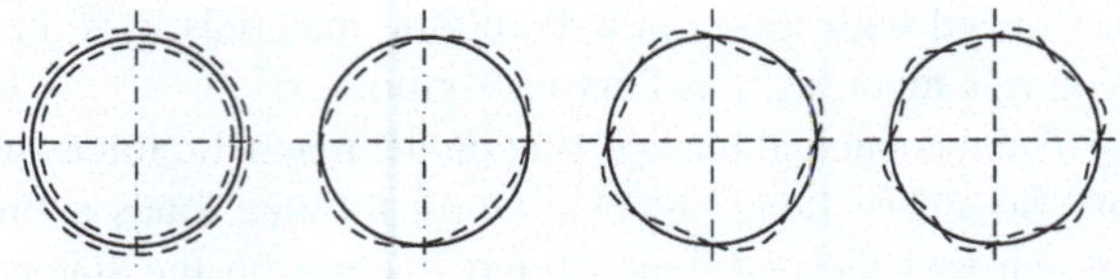

Fig. 1 Stator core oscillation modes

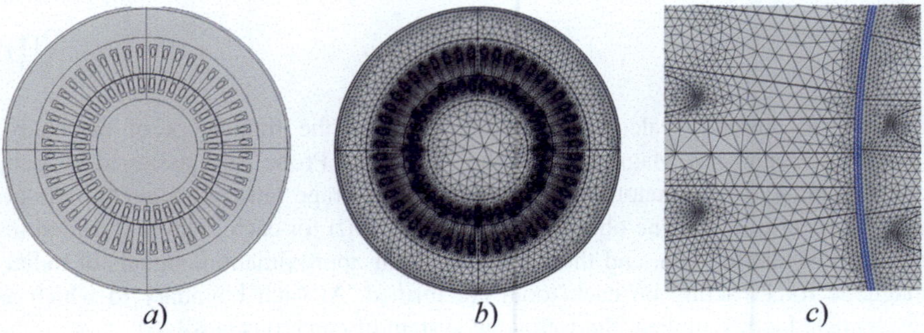

Fig. 2 Numerical model of the AIR180M4U3 engine: model geometry (**a**), model computational grid (**b**) and air gap (**c**)

3 Results

3.1 Geometry of the Model

The developed numerical model has a sufficiently high accuracy and high calculation speed due to the high grid density. Full rendering of the section of the AIR180M4U3 electric motor allows conducting research with asymmetric geometry. The geometry of the numerical model of the motor and its calculation grid are shown in Fig. 2.

Figure 2b, c shows that a high-quality triangular grid with a distinct structure and increased density in the air gap region was used in the calculations. The number of grid elements, depending on the eccentricity radius, varied from 129 thousand to 172 thousand.

3.2 Description of a Rotating Electric Machine

In the Rotating Domain module, it is necessary to specify the source of motion (kinematic, in which the rotor rotation is determined by an analytically specified law, or dynamic, in which the rotor rotation is determined by magnetic forces), as well as the magnitude of the load moment, the rotor moment of inertia, the initial speed, the phase and the axis of rotation. The Rotating Domain module solves problems associated with changing the geometry of a rotating engine, and the rotating region

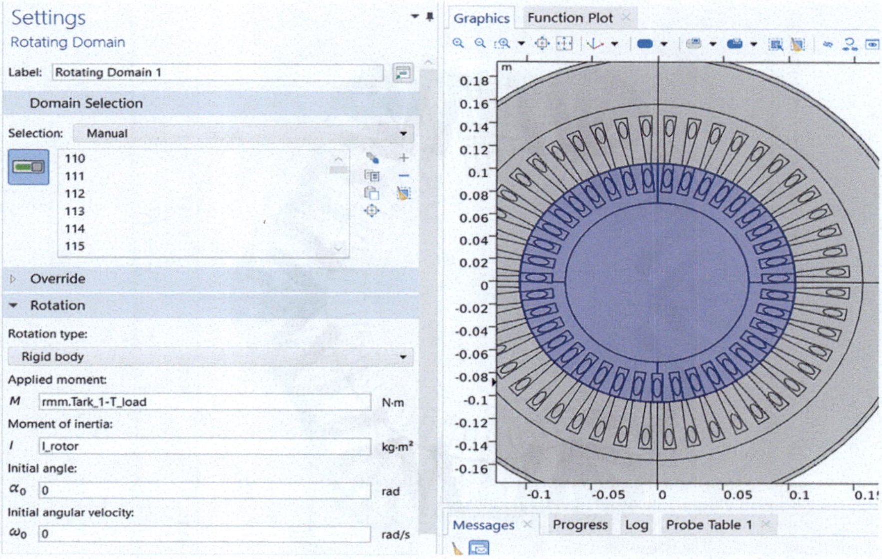

Fig. 3 Rotating Domain module

is selected as active—the rotor and the adjacent part of the air gap, as shown in the Fig. 3.

3.3 *Definition of Dynamic Parameters*

The rotor inertia moment was determined numerically using a special Mass Properties module. This function is especially useful when investigating motors with dynamic rotor eccentricity or with uneven load. The Mass Properties module settings are presented in more detail in Fig. 4.

The calculation of magnetic forces that drive the rotor is carried out using the "Force Calculation" block. Therefore, the rotor domains were selected for calculations within this block: its core and shaft.

In addition, the direction and position of the rotation axis and the point relative to which the torque calculation was performed were specified in this block, as shown in the figure. Here, the value of e_r corresponds to the radius of static eccentricity—the value by which the rotor axis is shifted relative to the stator axis (by default, equal to 0).

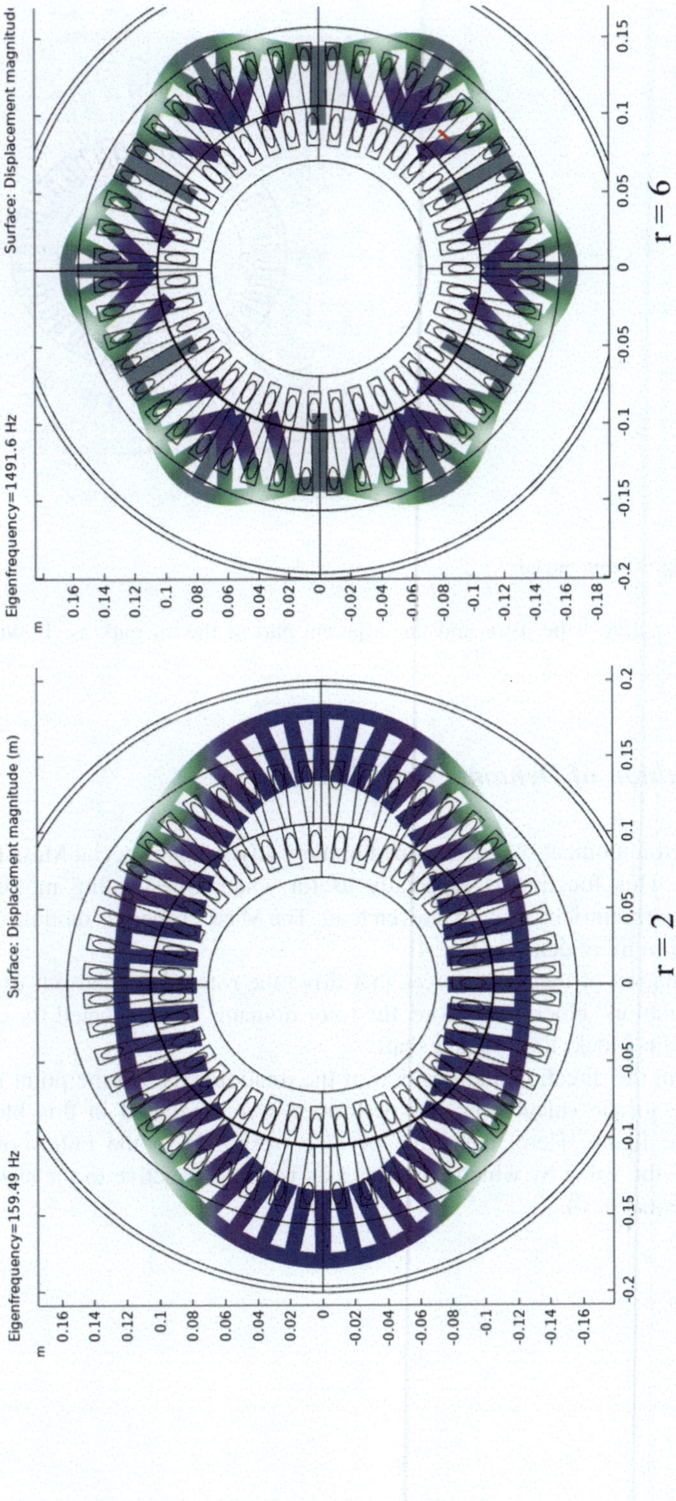

Fig. 4 Deformation of the rotor core at different forms of oscillations of the stator core

3.4 Determination of Frequency Spectra of Stator Core Deformation

Since the study is conducted in the frequency domain, we select the "Frequency Domain" block. To configure the block, you should select the initial and final excitation frequency of oscillations, as well as the frequency change step. Figure 4 shows the deformation of the rotor core for different forms of oscillation of the stator core. The frequency spectrum of the stator core deformation is shown in Fig. 5.

4 Discussion

The model allows the parameters and variables of all windings to be displayed in the form of a table. When studying in the time domain, it is possible to visualize this table in the form of a graph. The authors, in particular, obtained graphs of phase currents when starting an electric motor, diagrams of power losses in the phase windings of the motor and graphs of speed during starting, braking, and sudden changes in the load of electric machines.

The obtained graphs of magnetic induction in a rotating asynchronous motor indicate the continuity of magnetic lines as they pass through the air gap, which indicates the correct assignment of boundary conditions in this area.

The values of the electromagnetic moment obtained by calculating magnetic forces and using the Arkkio method are in good agreement (Fig. 6).

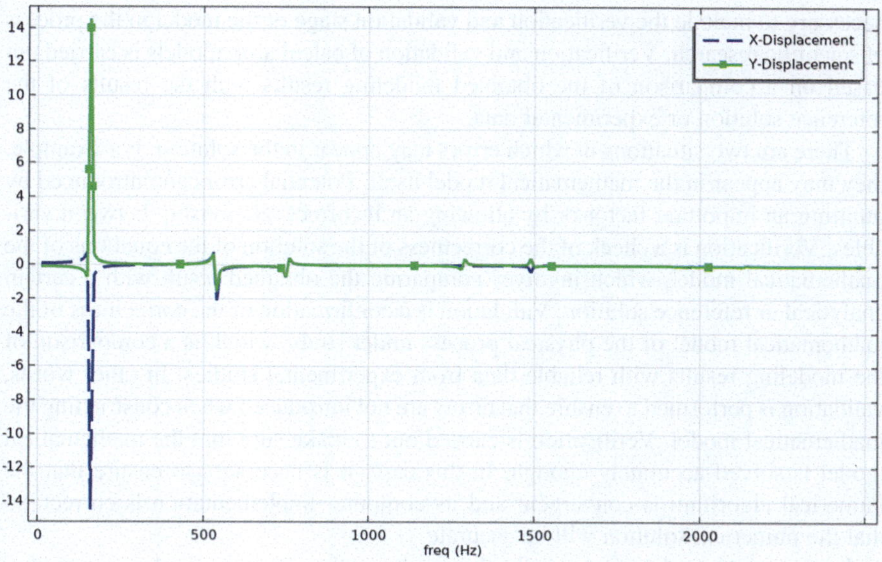

Fig. 5 The frequency spectrum of the stator core deformation

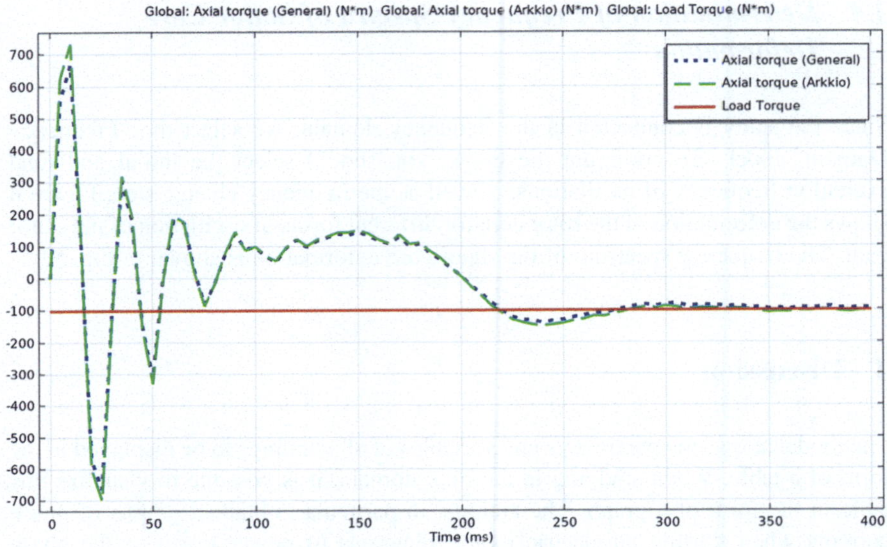

Fig. 6 Electromagnetic moment obtained by calculating magnetic forces and using the Arkkio method

In the process of developing a mathematical model, numerical modeling and processing the results, due to the selection of incorrect initial data, equations and boundary conditions, as well as assumptions, rounding of numbers and other subjective factors, errors creep into the calculation model and, as a consequence, into the final results. Thus, to check the correctness of the calculation model, it is necessary to include the verification and validation stage of the model in the process of scientific research. Verification and validation of calculation models is carried out based on a comparison of the obtained modeling results with the results of the reference solution or experimental data.

There are two situations in which errors may appear in the solution. For example, they may appear in the mathematical model itself. Potential errors are introduced by omitting an important factor or by allowing an incorrect relationship between variables. Verification is a check of the correctness of the solution of the equations of the mathematical model, which involves comparing the obtained result with a certain analytical or reference solution. Validation is a confirmation of the correctness of the mathematical model of the physical process under study, which is a comparison of the modeling results with reliable data from experimental studies. In other words, validation is performed to ensure that errors are not introduced when constructing the mathematical model. Verification is carried out to make sure that the mathematical model is solved accurately enough. In this case, it is necessary to ensure that the numerical algorithm is convergent and its computer implementation is correct, so that the numerical solution will be accurate.

In numerical modeling, to verify the results in the absence of reference mathematical solutions and experimental data, one can use the artificial solution method.

The goal of solving the initial-boundary value problem is to find an explicit expression for the solution in terms of independent variables, usually space and time, taking into account the specified parameters of the problem, such as material properties, boundary and initial conditions. The general form of the initial conditions implies a description of all forces acting on the body.

In the artificial solution method, the problem is solved "by inverse," i.e. it starts with a hypothetical explicit expression for the solution. The solution is then substituted into the differential equations to obtain a consistent set of initial, boundary and base conditions. This usually involves evaluating a number of derivatives.

Similarly, the hypothetical solution at time $t = 0$ and at the boundaries is calculated to obtain the initial and boundary conditions.

Verification is completed by the verification stage. Given the initial and auxiliary conditions just obtained, a modeling tool is used to obtain a numerical solution to the initial boundary value problem. The result obtained is compared with the initially hypothetical solution from which the numerical modeling began.

The combination of verification and validation is called qualification. This procedure is necessary because there is a possibility that experimental data will accidentally coincide with an incorrect solution as a result of a combination of an incorrect mathematical model and an incorrect solution algorithm. Except for such rare cases, a good agreement between the numerical solution and the experimental results confirms the validity of the mathematical model and the accuracy of the solution procedure. High mesh quality is a necessary condition for checking the correctness of the results in combination with a correct formulation of the problem, the use of correct boundary conditions and initial data.

5 Conclusion

A finite element model of an electric machine that allows one to study the effect of changes in the load moment on its shaft on the excitation of oscillations of magnetic fields in its air gap, allowing one to numerically estimate the values of these state variables. Using this model allows one to identify the conditions for the occurrence, propagation and concentration of oscillation power in the systems "prime engine—electric generator—power supply system load with a DC link" and "power grid—frequency converter of electric power—electric motor—working machine."

The algorithms for calculating the indicators used in this study allow us to estimate the influence of torsional vibrations, shaft beats (eccentricity) and spatial-temporal distributions of magnetic induction and radial magnetic force in the air gap of an electric machine, generating magnetic noise, on energy losses and the operating life of an electromechanical complex. The dependencies found will allow us to develop software for all stages of designing newly created vibration and noise protection systems for electric machines associated with such frequency converters as inverters and active rectifiers. The program allows us to calculate the dynamics of vibration powers and useful load—mechanical for the drive and electrical for the generator.

Acknowledgments The study was supported by the grant of the Russian Science Foundation No. 20-19-00372-П, https://rscf.ru/project/20-19-00372/.

References

1. Araújo, R.: Induction motors—modelling and control. IntechOpen (2012)
2. Vidhya, B., Srinivas, K.N.: Vibration analysis including stator, rotor, housing and dynamic response analysis of flux reversal generator. Journal of Electrical Systems and Information Technology. **5**(2), 144–157 (2018)
3. Zhao, G., Sun, L., Niu, N., Wang, X., Xing, Z.: Analysis of vibration characteristics of stators of electrical machines under actual boundary. Mech. Syst. Signal Process. **185**, 109778 (2023)
4. Xu, X., Han, Q., Chu, F.: Review of electromagnetic vibration in electrical machines. Energies. **11**, 1779 (2018)
5. Ermolaev, A., Erofeev, V., Plekhov, A., Titov, D.: Magnetic vibration in induction motor caused by supply voltage distortion. Energies. **15**(24), 9600 (2022)
6. Leleu, E., Espanet, C., Miraoui, A., Siala, S.: Analytical modelling of electromagnetic origin vibrations in an induction machine supplied by high power PWM inverter. In: In 32nd Annual Conference on IEEE Industrial Electronics, pp. 836–843. IEEE, Paris (2006)
7. Erofeev, V.I., Ermolaev, A.I., Plekhov, A.S., Titov, D.Y., Gorokhova, I.Y.: Study of magnetic vibration occurred in a squirrel-caged induction motor caused by distortion of the supply voltage waveform. Engineering Physics. **11**, 39–52 (2023)
8. Ibrahim, I., Silva, R., Lowther, D.A.: Application of surrogate models to the Multiphysics sizing of permanent magnet synchronous motors. IEEE Trans. Magn. **58**(9), 1–4 (2022)
9. Ling, X., Li, B., Gong, L., Huang, Y., Liu, C.: Simulation of switched reluctance motor drive system based on multi-physics modeling method. IEEE Access. **5**, 26184–26189 (2017)
10. Pyrhonen, J., Hrabovcova, V., Semken, R.S.: Electrical Machine Drives Control: An Introduction. Wiley, United Kingdom (2016)
11. Yin, Q., Wei, X., Nie, H., Deng, J.: Study on modal characteristics and vibration reduction of an aircraft rotor–stator brake-induced squeal system. Acta Mech. Sinica. **36**, 1350–1359 (2020)
12. Li, H., Da, Z., PeiYun, X., Cheng, C., DongMei, H., XiaoJun, Y., Zhihuan, S., Zhihong, H.: Analysis on the vibration modes of the electric vehicle motor stator. In: Vibroengineering PROCEDIA 22, pp. 81–86. JVE International, Dubai (2019)

Information-Analytical System for Fractal and Neural Network Diagnostics of Computed Tomography Images of the Lungs

Vladislav Salmiyanov and Anna Maslovskaya

Abstract This paper is devoted to the development and testing of a software package designed for intelligent prediagnosis of digitized computed tomography images of the human lung. A hybrid approach for combining algorithms of digital image processing and analysis is proposed for the development of the software application. In the first step, an algorithm for pre-processing digital images based on filtering and segmentation techniques has been presented. The second step involves the implementation of algorithms for the assessment of scaling characteristics using fractal and multifractal image analysis methods. The classification of computed tomography images was carried out using a convolutional neural network construction and training method. The characteristics of the Renyi spectra for images corresponding to the identified pathologies were analyzed. The classification of the images into the categories corresponding to 'norm' and 'pathology' on the training set was also performed using the developed techniques.

Keywords Computed tomography scan · Fractal analysis · Multifractal analysis · Renyi spectrum · Convolutional neural network · Image preprocessing

1 Introduction

The advent of computer technology has opened up new possibilities in many areas, including the medical practice of non-invasive diagnosis of the human respiratory system. Accurate diagnosis of the respiratory system at the earliest stages of disease is of paramount importance to ensure the success of subsequent treatment strategies. The selection of non-invasive diagnostic methods is based on the specific

V. Salmiyanov
Innopolis University, Innopolis, Russia

A. Maslovskaya (✉)
Innopolis University, Innopolis, Russia
e-mail: a.maslovskaya@innopolis.ru

A. Gibadullin, S. Gordin (eds.), *Computing Technologies and Applied Mathematics*, Springer Proceedings in Mathematics & Statistics 500,
https://doi.org/10.1007/978-3-031-90546-9_5

requirements of the examination. For example, chest X-rays (which produce images of the heart, lungs, blood vessels, airways and bones of the chest and spine) provide rapid results and have low radiation exposure. However, they have a relatively low resolution, which can lead to potential diagnostic errors and even missed pathology. Obviously, computed tomography (CT) of the lungs can significantly improve resolution and facilitate the detection of pathological changes that may not be visible on x-ray. The successful completion of diagnostic and classification tasks depends on the ability to identify and distinguish patterns. To achieve this, it is essential to analyse large amounts of data and detect specific patterns. In the context of pattern recognition, the challenge of analyzing complex systems arises from the non-linear nature of these systems, which makes them difficult to interpret.

The advent of modern computer technology has made it possible to develop sophisticated automated diagnostic systems. Such systems are designed to assist medical professionals in making accurate diagnoses and to minimize undetected human error, which can lead to misdiagnosis or the omission of pertinent details in the evolving morphology of the human body. In addition, the application of sophisticated image processing methods can significantly improve the accuracy of disease identification. Currently, a significant proportion of automated diagnostic software packages are based on data mining techniques, neural network training algorithms or a combination of both.

In addition, fractal theory has been used in many biomedical disciplines to elucidate the morphological structure of irregular patterns. A sophisticated method of fractal analysis is multifractal analysis, which allows not only the determination of the complexity of the structural pattern, but also the identification of scaling characteristics consistent with the morphological configuration of the object [1–3]. In particular, fractal theory has been widely used in the analysis and characterization of irregular biomedical entities. Fractal and multifractal theory has been the subject of considerable interest and has offered potential computational techniques for the analysis of the complexity and irregularity of events observed in biomedical images [4]. It is important to note that in a number of cases the assessment of the fractal dimension is not informative. In such cases, multifractal analysis can be used to provide a statistical assessment of the fractal characteristics and an estimate of the spectrum of dimensions. For example, in reference [5] the researchers have shown the correlations between changes in fractal dimension and dose variations for tumor growth in human lungs.

The utilization of fractal analysis in clinical practice must be substantiated, particularly insofar as it serves only as an ancillary technique. For example, the fractal dimension appears to be an oversimplified parameter that is not correlated with the spatial heterogeneity of radioaerosol distribution, as observed in [6]. A set of reference information features based on fractal analysis has been analyzed in [7] to create algorithms for diagnosing tuberculosis pathology. In regard to addressing medical challenges, the authors of [8] identified a correlation between fractal dimension and lacunarity as potential diagnostic biomarkers for small cell lung cancer. These biomarkers exhibit fluctuations in value at varying stages of treatment, as determined by computed tomography images. In [9], a tool for the early detection of lung pathologies caused by air pollution has been developed. In this process, the fractal dimension of CT image segments serves as the diagnostic tool. A quantitative assessment of chest CT images of patients with 2019 novel coronavirus disease (COVID-19) was conducted using multifractal algorithms, as reported in [10].

Researchers have identified a correlation between the multifractal spectrum and disease progression or regression. They also identified a threshold fractal dimension that predicts the likelihood of a patient's recovery. The proposed method has the potential to be a valuable tool for clinicians in the diagnosis and treatment of pneumonia caused by the novel coronavirus (2019-nCoV) and the timely identification of therapeutic effects. The proposed method has the potential to be an effective tool for physicians in the diagnosis and treatment of pneumonia caused by the novel coronavirus (2019-nCoV) and in the timely identification of the efficacy of therapeutic interventions. A system of multifractal data mining in lung CT imaging has been proposed in reference [11]. The analysis was performed using a preprocessing segmentation algorithm and the multifractal parameterization method implemented on the Matlab platform. Another promising alternative is the multifractal parameterization based on wavelet transforms [12].

Another notable advance has been the development of neural networks, which facilitate the collection and mining of an expanded range of data, including unstructured data. It is widely recognized that neural networks facilitate the solution of problems in a range of disciplines, including pattern recognition, image restoration, prediction, and so on. Approaches to the use of neural networks can be divided into two main groups: pre-trained models and the construction of custom networks. The advantages of pre-trained models include ease of use, as users only need to start training with their own dataset. In addition, the pre-trained weights are already set, allowing faster training and reducing the amount of data required to achieve optimal performance. However, there are some limitations to using pre-trained models, including limited flexibility and difficulty in adapting to new datasets, as the network has been trained on specific data. Another disadvantage is the possibility of performance degradation after retraining. The use of custom neural networks can overcome the limitations of pre-trained models, as the network is tailored to a specific problem.

In [13], for example, researchers attempted to use neural networks to reconstruct computed tomography images. In [14], the study focused on optimising image processing for magnetic resonance imaging (MRI) research. A neural network was used as an optimizer to reduce noise in the images, resulting in a significant reduction in the processing time for image analysis from 35 min to 6 min. The functionality of a software application for automated early cancer diagnosis has been documented in reference [15]. The authors integrated VGG-19 and long short-term memory (LSTM) neural networks with image segmentation algorithms, resulting in an improved detection accuracy of 98.8%. In addition, previous studies have addressed similar issues. For example, [16] investigates the effectiveness of classifying malignant and benign brain tumors using Python and MATLAB software. The methods of intelligent image analysis and classification using neural networks are dependent on the pre-processing algorithms used for image manipulation. Image manipulation can be broadly categorized into general image enhancement, such as the use of filters, contrast adjustment and morphological operations. These can be used to create algorithms for binary conversion and highlighting of areas. In essence, image manipulation provides the basis for the development of automated human diagnostic systems. It is noteworthy that the authors in [17–19] have used automated segmentation algorithms to detect signs of pathology in various human body systems.

The aim of the present study was to develop a software package based on a combination of multifractal techniques and neural networks, with the aim of exploring multidisciplinary possibilities for intelligent pre-diagnosis of digitized computed tomography images of the human lung. The study used a combination of real CT scans and images from an open access database [20] to develop the model. Two case studies were selected for this research, representing both normal patients and patients diagnosed with pneumonia. The paper also outlines the pre-processing techniques used to improve the accuracy of pathology detection.

2 Theoretical Foundations and Programming Tools for Image Analysis

2.1 Fractal Dimension Assessment

As mentioned above, the diagnosis of lung pathology based on fractal analysis of the self-similarity properties of raster images is an actively developing area of research. In general, a variety of techniques can be used to quantify the fractal dimension of two-dimensional biomedical objects. The selection of an appropriate measurement technique depends on the intrinsic nature of the fractal object and the range of scales over which the analyzed object can be considered to exhibit fractal characteristics. In order to determine the fractal dimension D, real space measurement techniques are typically used. These are based on the construction of various coverings, namely the use of the definition of the Hausdorff—Besikovich dimension or its direct consequences [21]. Various implementations of this method are known, but most of them are based on the use of the following "number of particles N – radius R_0" relationship

$$N \sim (R/R_0)^D. \quad (1)$$

The "box-counting" method is based on an algorithm that varies the particle sizes of the set covering the image. This method is the most widely employed technique for determining the fractal dimension. The principal advantages of this approach are the simplicity of the computational process and the relatively straightforward implementation on a computer. However, the method has two main disadvantages. Firstly, it is only possible to calculate the fractal dimension from binarized images, and secondly, it is not possible to compute from color or grayscale images. The box-counting algorithm for calculating fractal dimension is based on covering a digitized and binary-encoded image with square grids. The grid number N with a specific linear size R_0, that contains at least one pixel from the original image, is then determined. Further, the dependence of the number of squares N covering the cluster on their size $\log N \sim -D \log R_0$ is plotted using a double logarithmic scale. Next, the slope coefficient of the linear regression line, which approximates the results of the data analysis, allows one to calculate the fractal dimension D.

The fractal dimension of a raster image can be calculated in two ways: by analyzing the clusters of boundaries, and by analyzing the image as a whole. Although the methods used to determine these two dimensions are analogous, there is a crucial difference. The fractal dimension of the boundaries indicates the degree of indention of an object, whereas the fractal dimension of the image structure indicates the extent to which the object occupies space.

2.2 Multifractal Analysis

Furthermore, multifractal analysis offers a range of techniques that can be applied in a variety of ways. The defining characteristic of multifractals is their heterogeneous structure, which precludes the possibility of describing them using a single fractal dimension. In order to fully characterize the structure of an object, it is necessary to use a range of dimensions. One of the most widely used algorithms for determining multifractal characteristics can be considered as an extension of the classical "box-counting" method [21].

The first step of the multifractal parameterization technique involves covering the binary image (defined as the set $\Psi_{N \times N}$) by square clusters with variable cluster size l. Next, let us assume that the set Ψ consisting of $N \times N$ points has a certain number of points in the (i,j) cluster, $i,j=\overline{1,k},\ \ k=N/l$, and then create the matrix C_{ij} with each cell containing the number of pixels from the studied image falling within the scanning cluster. After that, the specific weight of each cluster is calculated by dividing the number of pixels within that cluster by the total number of pixels in the entire image:

$$P_{i,j} = \frac{C_{ij}}{\sum\limits_{i,j} C_{ij}}, \tag{2}$$

where $\sum\limits_{i,j} C_{ij}$ is the total number of units in the cluster matrix.

Then, the measure for the fractal dimension d is calculated as follows:

$$M_d(q,l) = \sum_{i,j=1}^{k} P_{ij}^{q} \cdot l^d = N(q,l) \cdot l^d \underset{l \to 0}{\to} \begin{cases} 0, & d > \tau(q) \\ 0, & d > \tau(q) \end{cases}, \tag{3}$$

where l is the cell size; q is the order of the moment.

The measure is characterized by a sequence of indicators $\tau(q)$ that determine how the probabilities change according to a power law depending on l. From formula (3) it follows that the weighted number of cells can be written in the form:

$$N(q,l) = \sum_{i,j}^{k} P_{ij}^{q} \propto l^{-\tau(q)}, \tag{4}$$

and the mass index is determined by:

$$\tau(q) = -\lim_{l \to 0} \left(\frac{\ln N(q,l)}{\ln l} \right) \tag{5}$$

Hence, using formulas (4) and (5), we can calculate the weighted number of cells $N(q,l)$ and the scaling exponent $\tau(q)$ under variation of the deformation parameter q. Finally, the Renyi spectrum of fractal dimensions is determined by the following expression:

$$D(q) = \frac{\tau(q)}{q-1}. \tag{6}$$

The key tools for image analysis using the multifractal parameterization method are the Renyi dimension range and the scaling exponent. These tools allow us to determine the characteristics of the studied structure. In addition, the zeroth moment indicates the fractal dimension, which is determined by the coating technique.

2.3 A Neural Network Technique for Image Processing

In addition, the present study assumes the use of software tools for the intelligent classification of raster images. The analysis focuses on the creation and training of a convolutional neural network. The Adam optimizer was used as the training algorithm for the neural networks. In this way, the results of training several neural networks using different optimizers allowed us to show that the Adam optimizer achieves the highest accuracy. Training was carried out for 40 iterations. In the next process, images with dimensions of 224 × 224 pixels were fed into the neural network. After transformation by layers of the neural network, the image data is processed by control layers. The design of the network architecture includes the following features: initially there are no control layers, and the first convolutional filter in the first layer has a size of 7 × 7 pixels. Subsequent convolutional layers use filters with a size of 2 × 2 pixels. This design choice is based on the idea that the first convolutional layer captures more detailed information in the image, leading to improved accuracy during training. Figure 1 shows the structure of the neural network and its layer parameters.

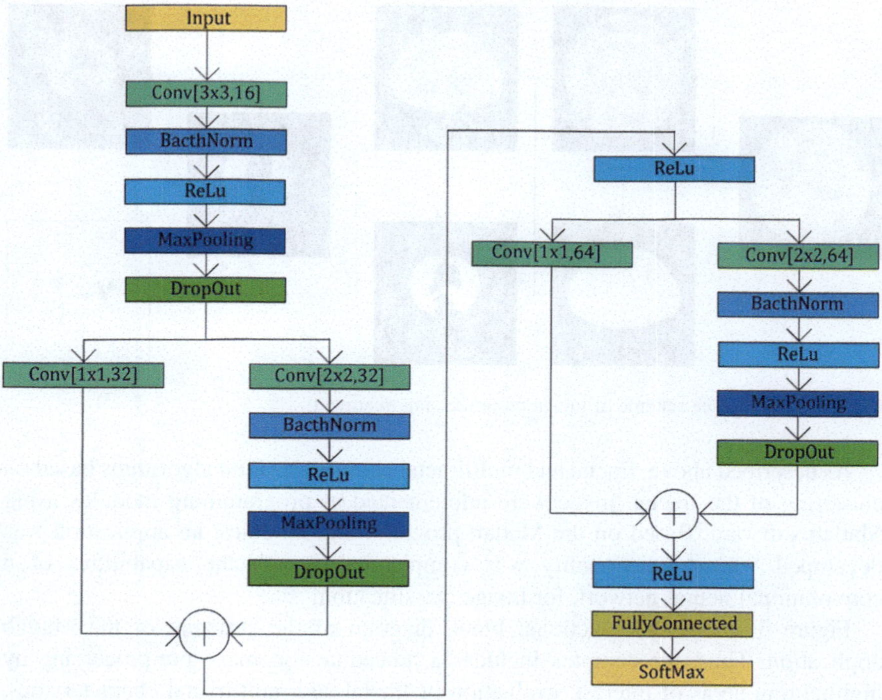

Fig. 1 The structure and configuration of a neural network

2.4 *Image Preprocessing and Software Application Functionality*

Original medical scans stored in DICOM format contain CT lung images. To speed up the computations, we converted the source images from DICOM to PNG format using subsequent image processing. The next step in the process requires an adaptive segmentation algorithm. We have divided this algorithm into a series of transformations. First, a bitmap image is loaded. Then an average background is set based on the original image. This allows us to automatically select the segmentation coefficients needed to create a binary mask that removes objects unrelated to the patient. Next, we repeat the process of averaging the background and creating the binary mask on the transformed image. In this case, however, we focus on highlighting the human lung as the object of study.

In the second stage of the segmentation process, a transformed image of the human lung is loaded. This image has been segmented and the 'bottom of hat' transformation has been applied. This transformation increases the contrast of the image and facilitates a more accurate binary classification of the objects to be further analyzed. The steps of the adapted segmentation process are illustrated in Fig. 2 below.

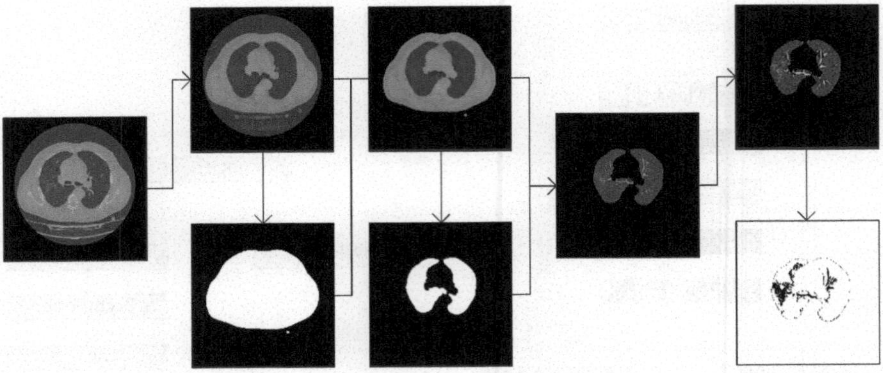

Fig. 2 The sequence scheme of image preprocessing treatment

As described above, fractal and multifractal parameterization algorithms based on clustering of the fractal image were implemented as programming modules using Matlab software. Based on the Matlab programming modules, an application was developed whose functionality was complemented with the capabilities of a convolutional neural network for image classification.

Figure 3 shows the functional block diagram of the structure of the Matlab application. Thus, the complex includes a procedure for image pre-processing by highlighting areas of interest, evaluation of fractal and multifractal characteristics, and image classification based on the neural network approach.

3 Results and Discussion

In view of the above, we would like to present the results of intelligent image analysis using the developed system, which includes the specification of the fractal dimension and multifractal properties, as well as the image classification tool based on the application of a convolutional neural network.

3.1 Scaling Characteristics for Lung CT Images

Following the principles and sequence of analysis stages implemented in the application of the programme, we will demonstrate the results of the study of images from CT scans of the lungs. To carry out the study, both normal and pathological images were selected, located in the same area of the lung. The CT scans were initialized in the Digital Imaging and Communications in Medicine (DICOM) format so that adaptive segmentation algorithms could be used to isolate the human lungs and then binaries the images.

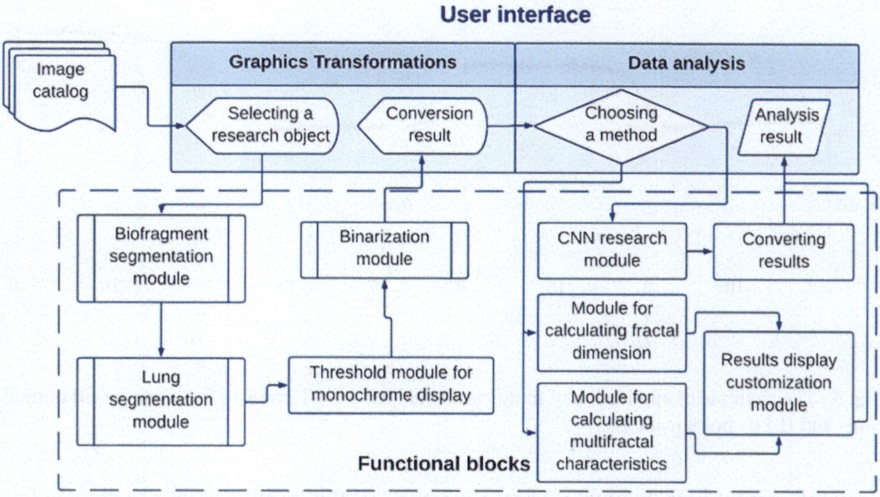

Fig. 3 The functional block diagram for Matlab program application

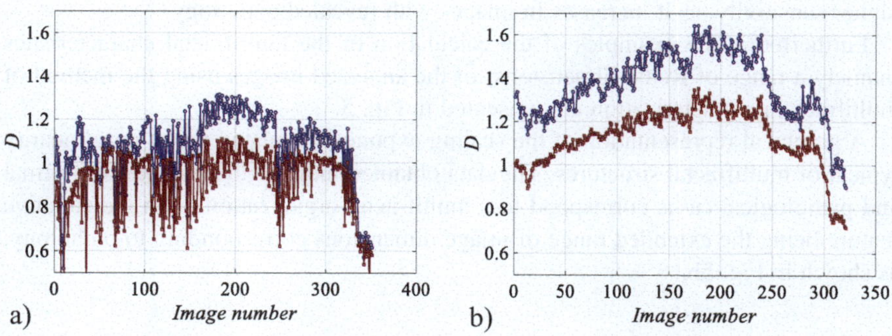

Fig. 4 The example of application of fractal analysis to a set of CT images of lungs for (**a**) normal and (**b**) pathological cases

The fractal dimensions of the images were calculated along the boundaries of the fractal clusters as well as for the entire image structure. Since CT scans are a series of images obtained from a layered scan of the human body, the results of the calculations are presented as a graph, where each image corresponds to a value of the fractal dimension for the boundaries of the clusters and for the space occupancy. In addition, multiscale fractal parameterization was applied to selected images. Fractal analysis was applied to a set of segmented lung computed tomography scans. Figure 4 shows graphs of the analysis results for the image set, both for normal and pathological cases. In the graphs, the fractal dimension of the boundary is shown in red, while the fractal dimension of the internal structure is shown in blue.

The results obtained suggest that using only the numerical estimation of the fractal dimension of boundaries for normal and pathological cases cannot provide

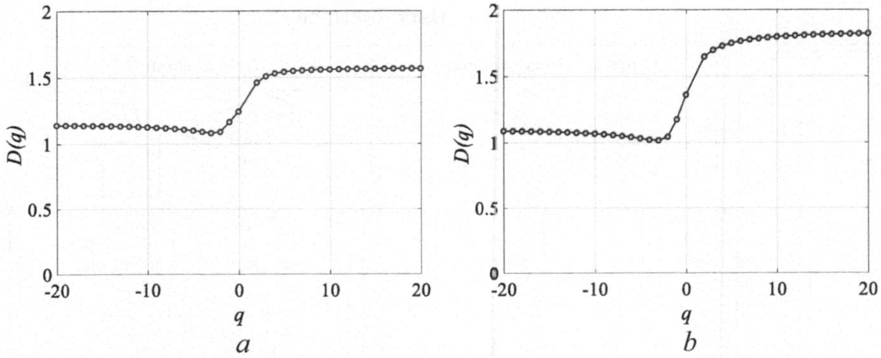

Fig. 5 The example of application of multifractal analysis for CT scans of the lungs for (**a**) normal cases and (**b**) pathological cases

a reliable method for diagnosis, since the values of these dimensions are in a similar range and vary depending on the slice scanned. On the other hand, the technique based on the estimation of the fractal dimension of the image structure shows a higher sensitivity, as it increases in images with revealed pathology.

Furthermore, the examples of the calculation of the multifractal characteristics (namely a range of Renyi dimensions) of the analyzed images using the method of multifractal parameterization are presented in Fig. 5.

A graphical representation of the scaling exponent is related to the non-linearity typical of multifractal structures. The data obtained indicate that images for normal and pathological cases correspond to a multifractal organization, with a distinctive feature being the extended range of image dimensions corresponding to pathology, as shown in Fig. 5b.

3.2 CT Image Classification

In the current study, we focused on the application of a convolutional neural network to provide classification of digital images of CT lung scans in the simplest case to solve the problem of classification of pathological and normal cases. CT scans were used to train the convolutional neural network and to test the classifier. The difference is that the images used for training were not subjected to morphological processing. Two diagnostic cases, one with normal and one with pathological findings, were selected for training.

The total number of images used was 14,286, with 6793 corresponding to normal findings and 7493 corresponding to pathological findings. Among this sample, 75% were used as a training sample, while the remaining 25% formed a test sample. The preprocessing algorithms involved normalizing the input data by formatting the images into monochrome and converting them to a resolution of 224 × 224 pixels.

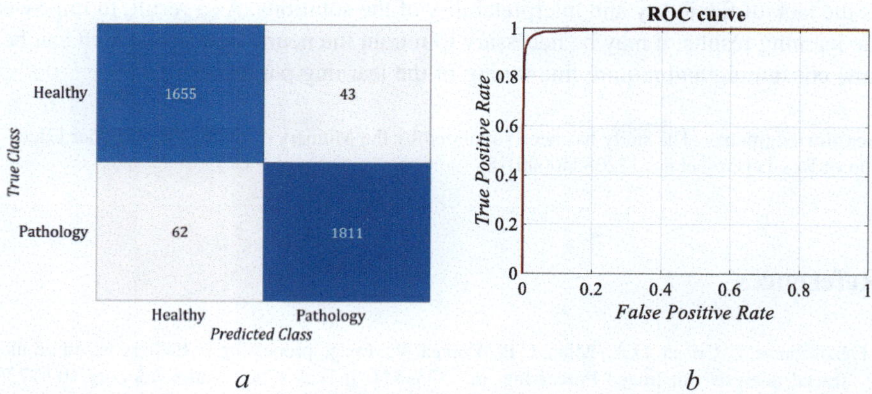

Fig. 6 Neural network learning metrics: (**a**) the error matrix and (**b**) the ROC curve

The results of the neural network learning process produced an error matrix and an ROC curve as shown in Fig. 6.

The error matrix (Fig. 6a) was generated using a test sample of images, with a total number of images corresponding to 3571 normal and pathological images. Based on this error matrix, the overall accuracy of the neural network was calculated to be 98.7%. The ROC curve (Fig. 6b) provides a more accurate characterization of network learning than the accuracy derived from the error matrix. A key feature derived from the ROC curve is the area under the diagonal. Therefore, the larger the area under this line, the more accurate the classification results of the network. Training on a set of CT scans resulted in an area under the ROC of 0.9953.

4 Conclusion

Thus, the paper summarizes an investigation of the designed program application based on a combination of fractal and multi-fractal analysis techniques, as well as neural network learning, for the analysis of CT lung images. Focusing on the analysis of CT images using fractal theory, pathology classification can be achieved by applying these techniques to a set of segmented images. However, sensitivity is only demonstrated when the fractal dimension of spatial occupancy is considered. When lung images are examined using fractal boundary analysis, the method does not show sufficient sensitivity due to the proximity of normal and pathological values.

In addition, multifractal parameterization has shown sensitivity in distinguishing between normal and pathological images, making it a useful tool for initial diagnosis of respiratory organ conditions. The use of neural network based learning has several advantages, a major one being the ability to perform batch classification of images without the need for pre-processing. However, there are also some drawbacks, such as the lack of flexibility and interpretability of the solutions. As a result, to improve the learning results, it may be necessary to retrain the neural network, which can be time consuming and require fine-tuning of the learning parameters.

Acknowledgments The study has been supported by the Ministry of Science and Higher Education of Russia (Project no. 122082400001-8).

References

1. Sztojonov, I., Crisan, D.A., Mina, C.P., Voinea, V.: Image processing in biology based on the fractal analysis. In: Image Processing, pp. 323–344. InTech (2009). https://doi.org/10.5772/7059
2. Reljin, I.S., Reljin, B.D.: Fractal geometry and multifractals in analyzing and processing medical data and images. Arch. Oncol. **10**(4), 283–293 (2002). https://doi.org/10.2298/AOO0204283R
3. Lopes, R., Betrouni, N.: Fractal and multifractal analysis: a review. Med. Image Anal. **13**(4), 634–649 (2009). https://doi.org/10.1016/j.media.2009.05.003
4. Lin, P.L., Huang, P.W., Lee, C.H., Wu, M.T.: Automatic classification for solitary pulmonary nodule in CT image by fractal analysis based on fractional Brownian motion model. Pattern Recogn. **46**(12), 3279–3287 (2013). https://doi.org/10.1016/j.patcog.2013.06.017
5. Chakravarty, S.P., Chakraborty, S.: Fractal analysis related to tumour growth in lungs: a review. In: Fourth International Conference on Image Information Processing, pp. 1–4, Shimla, India (2017). https://doi.org/10.1109/ICIIP.2017.8313752
6. Chung, H.-W., Huang, Y.-H.: Fractal analysis of nuclear medicine images for the diagnosis of pulmonary emphysema. Am. J. Roentgenol. **174**(4), 1055–1059 (2000)
7. Abdikerimova, G., Shekerbek, A., Tulenbayev, M., Sultanova, B., Beglerova, S., Dzhaulybaeva, E., Zhumakanova, K., Rysbekkyzy, B.: Detection of lung pathology using the fractal method. Int. J. Electr. Comp. Eng. **13**(6), 6778–6786 (2023)
8. Mamdetsariev, I., Mirzapoiazova, T., Lennon, F., Jolly, M.K., Li, H., Nasser, M.W., Vora, L., Kulkarni, P., Batra, S.K., Salgia, R.: Small cell lung cancer therapeutic responses through fractal measurements: from radiology to mitochondrial biology. Clin. Med. **8**(7), 1038 (2019). https://doi.org/10.3390/jcm8071038
9. Szigeti, K., Szabó, T., Korom, C., Czibak, I., Horváth, I., Veres, D.S., Gyöngyi, Z., Karlinger, K., Bergmann, R., Pócsik, M., Budán, F., Máthé, D.: Radiomics-based differentiation of lung disease models generated by polluted air based on X-ray computed tomography data. BMC Med. Imaging. **16**(14) (2016). https://doi.org/10.1186/s12880-016-0118-z
10. Astinchap, B., Ghanbaripour, H., Amuzgar, R.: Multifractal analysis of chest CT images of patients with the 2019 novel coronavirus disease (COVID-19). Chaos, Solitons Fractals. **156**, 111820 (2022). https://doi.org/10.1016/j.chaos.2022.111820
11. Salmiyanov, V.O., Maslovskaya, A.G.: Software package for segmentation and multifractal diagnostics system of lung computed tomography digital images. Vestnik Tomskogo Gosudarstvennogo Universiteta—Upravlenie, Vychislitel'naya Tekhnika i Informatika. **65**, 105–115 (2023)
12. Maslovskaya, A.G., Afanasov, L.S.: Algorithms of multifractal wavelet analysis in problems of specifying raster images of self-similar structures. Vestnik Tomskogo Gosudarstvennogo Universiteta—Upravlenie, Vychislitel'naya Tekhnika i Informatika. **53**, 61–71 (2020)
13. Ivanenko, M., Smolik, W.T., Wanta, D., Midura, M., Wróblewski, P., Hou, X., Yan, X.: Image reconstruction using supervised learning in wearable electrical impedance tomography of the thorax. Sensors. **23**(18) (2023). https://doi.org/10.3390/s23187774

14. Truszkiewicz, A., Bartusik-Aebisher, D., Wojtas, Ł., Cieślar, G., Kawczyk-Krupka, A., Aebisher, D.: Neural network in the analysis of the MR signal as an image segmentation tool for the determination of T1 and T2 relaxation times with application to cancer cell culture. Int. J. Mol. Sci. **24**(2) (2023). https://doi.org/10.3390/ijms24021554
15. Alsheikhy, A.A., Said, Y., Shawly, T., Alzahrani, A.K., Lahza, H.: A CAD system for lung cancer detection using hybrid deep learning techniques. Diagnostics. **13**(6) (2023). https://doi.org/10.3390/diagnostics13061174
16. Yilmaz, V.S., Akdag, M., Dalveren, Y., Doruk, R.O., Kara, A., Soylu, A.: Investigating the impact of two major programming environments on the accuracy of deep learning-based glioma detection from MRI images. Diagnostics. **13**(4) (2023). https://doi.org/10.3390/diagnostics13040651
17. Halder, A., Chatterjee, S., Dey, D., Kolec, S., Munshi, S.: An adaptive morphology based segmentation technique for lung nodule detection in thoracic CT image. Comput. Methods Prog. Biomed. **197**, 105720 (2020). https://doi.org/10.1016/j.cmpb.2020.105720
18. Singh, D., Singh, A.K.: Role of image thermography in early breast cancer detection- past, present and future. Comput. Methods Prog. Biomed. **183**, 105074 (2020). https://doi.org/10.1016/j.cmpb.2019.105074
19. Gharleghi, R., Chena, N., Sowmya, A., Beier, S.: Towards automated coronary artery segmentation: a systematic review. Comput. Methods Prog. Biomed. **225**, 107015 (2022). https://doi.org/10.1016/j.cmpb.2022.107015
20. Kaggle. https://www.kaggle.com/datasets/tawsifurrahman/covid19-radiography-database. Accessed 3 Jan 2024
21. Falconer, K.J.: Fractal Geometry: Mathematical Foundations and Applications, 3rd edn. Wiley, Chichester (2014)

Optimizing Resources in Information and Telecommunication Systems on the Base of Mathematical Simulation

Andrey Preobrazhenskiy, Tatyana Avetisyan, Yuriy Klimenko, and Yuriy Preobrazhenskiy

Abstract The paper considers the possibilities of mathematical simulation for resource optimization in modern information and telecommunication systems. The features of the mathematical model associated with resources in the information and telecommunication system, in which there can be several channels of information movement, are analyzed. In order to support transmission through different channels, the use of an appropriate system is required. The possibilities for taking into account various constraints are considered: on the intensity of information received in various departments of the organization, on the components that are part of the resource support system. It is investigated how resources will be maintained. The mathematical model for optimizing material resources is proposed. It is proposed to use the following restrictions: restrictions on the amount of information transmitted, restrictions in terms of the maximum speed of computing modules in the system, restrictions on the resources that are necessary in order to transfer the necessary volumes of information arrays, restrictions on the time of work of specialists in the information and telecommunication system, restrictions on the information that is necessary to ensure management processes. The solution to the management problem has been developed, taking into account the optimization of information resources.

Keywords Information system · Management · Resource · Model · Optimization

1 Introduction

It can be observed how information and telecommunication systems that are used in various fields are currently developing. To analyze them, it is necessary to rely on the methods of a systematic approach. Then it is possible to analyze technical, informational, economic and other parameters within the framework of unified methods. In

A. Preobrazhenskiy (✉) · T. Avetisyan · Y. Klimenko · Y. Preobrazhenskiy
Voronezh Institute of High Technologies, Voronezh, Russian Federation

A. Gibadullin, S. Gordin (eds.), *Computing Technologies and Applied Mathematics*, Springer Proceedings in Mathematics & Statistics 500,
https://doi.org/10.1007/978-3-031-90546-9_6

practice, it is necessary to improve the efficiency of information and telecommunication systems. To do this, you need to apply optimization methods and use various resources. Resources are used through different channels. In this regard, it is necessary to form a generalized methodology, on the basis of which it is possible to assess the effectiveness and quality of information systems. The aim of the paper is to develop a methodology for optimizing resources in information and telecommunication systems.

2 Materials and Methods

When considering the information and telecommunication system, it is necessary to take into account the structure of resource use (Fig. 1). At the same time, they come through several channels [1]. It is necessary to rely on the relation of non-empty sets when describing the resource support system.

$$S \subset R \times H$$

the designations S are taken into account—the set that corresponds to the options for the formation of the resource support system; R is the set that determines the set of resources; H is a set that determines the set of controls.

We will describe the following for a set of resources accordingly:

$$R \subset \times \{r_j : j \in J\},$$

where $r_j = \overline{1, R_j}$—correspond to the set of resources that are of the j-th type;

j—indexes for resource numbering;

×—denotes that the Cartesian product is realized;

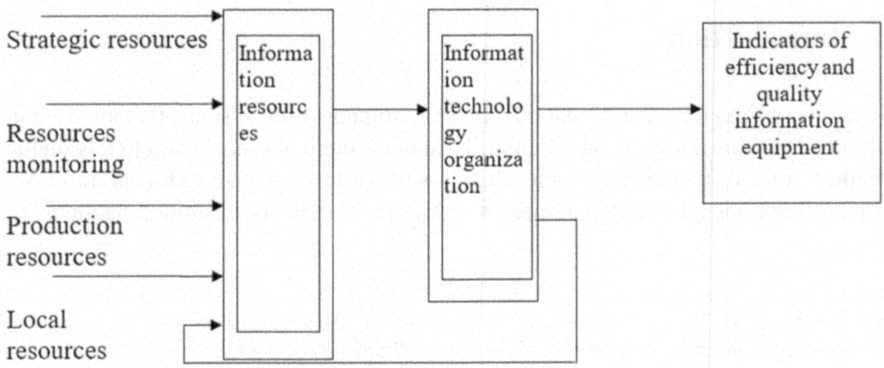

Fig. 1 The structure of the resource support system

$$H \subset \times \{h_t \div t \in T\}$$

where $h_t = \overline{1, H_t}$ – the set of directorates that have the t-th type;

T—multiple governance-related indexes.

The degree of training of specialists, the use of new technologies have a great impact on how resources will be used [2]. The features of the operation of the components of information and telecommunication systems depend on the choice of a set of appropriate departments. It is necessary to consider the corresponding vectors in order to represent the elements of the set (1)

$$S_l = (r, h), S_z \in S, l = 1, L$$

Herewith $r = (r_1, \ldots, r_j, \ldots, r_J)$, $j = 1,$, $r_j = 1, R_j$,

$$h = (h_1, \ldots, h_t, \ldots, h_T), t = 1, T, h_t = 1, H_t.$$

If a pair of elements will be represented like this $(r_j, h_t) = U_g$, $g = 1, G$, then

$$S_l = (U_1, \ldots, U_g, \ldots, U_G), l = 1, L.$$

it is necessary to take into account that $U_g = 1, U_g$, $g = 1, G$—elements of the resource support system,

L—the number of options for building a resource support system.

Requires the introduction of a vector of indicators f_{ug} in order to describe the elements of the resource support system U_g. For example, if management and external resources are considered, the number of detected malfunctions in information and telecommunication systems, the number of replacements in equipment [3] elements, etc., can be considered as an indicator. What the f_{ug} indicators will be affects the dependence of the integrated indicators of the functioning of information and telecommunication systems F_i, $i = \overline{1, I}$ while

$$F_i = \Psi(f_{ug}.). \tag{1}$$

For example, there is a dependence of the number of replacements in equipment elements within the entire information and telecommunication system on the values of such an indicator for all components of the resource management system. For example, there is a dependence of the number of replacements in equipment elements within the entire information and telecommunication system on the values of such an indicator for all components of the resource management system [4]. Such indicators can be combined into a subset $I_R \in I$. The operation of various components in the information and telecommunication system is described on the basis of other indicators, according to them (work with various information elements, preventive maintenance, equipment adjustment, etc.). The $I_H \in I$ subset is formed from the following indicators.

We can talk about such a relationship between the I_R and I_H subsets if, when the information and telecommunication system is managed:

$$I_R \cup I_H = I, I_R \cap I_H \neq \varnothing.$$

For sets U_g, there is the following relation of indicators:

$$f_{ug} \in \{U_{g \times} f\}, \Psi_i \in \{sxF_i : i \in I\}, \quad (2)$$

on their basis, it is possible to calculate values taking into account a certain combination of s elements U_g, while $F_i(s(U_g), f_{ug})$.

In the course of monitoring the information and telecommunication system, statistical information is collected on its various elements [5] in order to create dependencies (2). Then the random variables of F_{ug} are considered. This determines that there will be stochastic characteristics for $F_i = \psi_i(F_{ug})$ indicators. In this regard, it is necessary to determine probabilistic measures for them:

$$P\{\psi_s \subset (r, V) \times F_i \leq F_i^0 : i \in I\}, \quad (3)$$

in this case, $P\{\}$ determines the probability;

F^0_i is the value of the indicator F_i, which is necessary in order to solve the problem.

There are certain resources in the information and telecommunication system $r_g^0 \in U_g$, $g = \overline{1, G}$ makes it possible to determine h_g^0 so that there is a correspondence to the expression

$$P\left(\Psi_i C\left(r^0, h\right) \times F_i \geq F_i^0 : i \subset I_H\right) \rightarrow \max. \quad (4)$$

This rule applies when managing information and telecommunication systems in the course of selecting elements r_g^0, h_g^0:

$$P\left(\Psi_i C(r, h) * F_i \geq F_i^0 : i \in I_R \cup I_V\right) \rightarrow \max. \quad (5)$$

Given the set of relations (2), the probability of choice relative to the exponents $F_i(i \in I_R)$ is thus represented as follows

$$P(R) = \max P\{\Psi_i \in (r, h) \times F_i \geq F_i^0 : i \in I_R;$$

Given the set of ratios (3) on the indicators $F_i(i \in I_H)$, probability is represented by

$$P(H) = \max P\{\Psi_i \in (r, h) \times F_i \geq F_i^0 : i \in I_H.$$

Within condition (4), the probability associated with the choice will be p^{mp}, if condition (5), then the probability value will be p^c. Then, taking into account the construction of I_R I_H, subsets, for probability we write

$$p^{mp} = \frac{P(R,H)}{P(R)}; P^c = P(R) + P(H) - P(R,H),$$

then $P(R,H) = \max P\{\Psi_i \subset (r,h) \times F_i \geq F_i^0 : i \in I_R;$

$$\Psi_i \subset (r,h) \times F_i \geq F_i^0 : i \in I_H$$

In this case,

$$\mathrm{p^c - p^{\text{тр}} = p(R)(1 - p(H/R)) + (p(H) - p(H/R))} \quad (6)$$

We can say that the first term is always greater than 0 for (6). For the second, an assessment is required.

If we consider the values of $r_g^{\,0}$ that have been pre-selected, then the probability of correspondence taking into account the $F_i(i \in I_R \cap I_H$ and $h_g^{\,0}$ indicators) will be less than with r_g, which leads to the inequality $P(H) > P(H/R)$. That is, the second term and in general (6) will be positive.

Several typical structures can be combined, taking into account the control and resource components in the overall multi-channel system [6]. At the same time, there can be three types of such systems: convergent (first type), divergent (second type) and reverse (third type) (Fig. 1).

In the first type, the control and resource components are connected in the following way:

$$U_{sh} = \left(r_1^{\,inp}, \ldots rj^{inp}, \ldots r_7^{\,inp}, h_{sh}\right),$$

In the second type—in this way

$$U_{rasch} = \left(r, h_{rasch} r_1^{\,inp}, \ldots rj^{inp}, \ldots r_7^{\,inp}\right),$$

In the third type—in this way

$U_{rey} = (r, h_{rasch}, r_R^{\,out}, r_R^{\,inp})$. For U_{sh}, U_{rasch}, and U_{rey}, it is necessary to make the appropriate choice taking into account the optimization methods, as well as the F_i $(i = 1,I)$ indicators (Fig. 2).

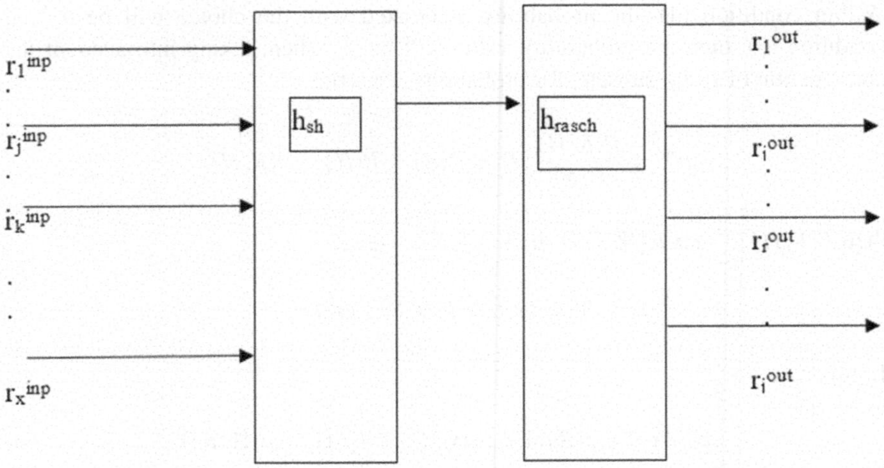

Fig. 2 Model of an information and telecommunication system in which several channels are used

3 Methodology for Optimizing Material Resources

If the distribution of resources is considered within a multi-channel information system, then it is possible to rely on the first and second types of structure. The second type of structure is considered to be the main one, which has an impact on management processes. The first type in the course of resource allocation is related to their sources. Information and telecommunication systems can be used to ensure the achievement of various goals in organizations, while it is important [7] to perform different types of work. For each of the j-th type of work in the system, it is necessary to determine the priority. In this case, a rational allocation can be determined from the existing resources $x_j \left(j=\overline{1,J}\right)$. Among the various methods of how priorities will be found, we can note the use of a priori ranking in expert assessment.

We can represent the absolute priorities $a_j \left(j=\overline{1,J}\right)$ found in a relative way:

$$c_j = 1 - \frac{a_j}{\sum_{i=1}^{J} a_i}, \quad j=\overline{1,J},$$

We introduce the objective function in the problem of optimal resource allocation as follows:

$$\sum_{j=1}^{J} c_j x_j \to \max. \tag{7}$$

It is also required to take into account the following restrictions in the task:

(a) by the intensity of information received by various departments [8] of the organization

$$\sum_{j=1}^{J} d_{kj}x_j \leq D_k, k=\overline{1,K}, \tag{8}$$

here d_{kj} is the average intensity of information inflow to the k-th department of the organization;

D_k is the total time of operation of the k-th department;

(b) for the components that are part of the resource support system:

$$\sum_{j=1}^{J} b_{gj}x_j \leq B_g, g=\overline{1,G}, \tag{9}$$

here, if we consider the j-th type of work in the organization under the g-th resource, there will be a coefficient of utilization of b_{gj};

B_g—planned support for the g-th resource.

Since information $\widetilde{D}_k, k=\overline{1,K}$ and resources $B_g, g=\overline{1,G}$ will move randomly within the organization, then a probabilistic description should be used to record constraints (8), (9):

$$P\left(\sum_{j=1}^{J} d_{kj}x_j \leq \widetilde{D}_k\right) \geq P_k^*, k=\overline{1,K}, \tag{10}$$

$$P\left(\sum_{j=1}^{J} b_{gj}x_j \leq \widetilde{B}_g\right) \geq P_g^*, g=\overline{1,G}. \tag{11}$$

It is important to note that the limits will be valid with a suitably defined probability P^*.

From the probabilistic representation (10), (11) it is possible to proceed to the deterministic (8), (9) representation, if there is information about their distribution laws for random variables. Within the organization, it is necessary to provide a resource plan for the types of work $x_j^{nn}, j=\overline{1,J}$. For such resources [9], alternatives with a total number m of options can be matched:

$$x_{jm}^{nn}, m=\overline{1,M}, j=\overline{1,J}$$

In addition, there are corresponding restrictions

$$x_j \geq x_{jm}^{nn}, j=\overline{1,J}. \tag{12}$$

Alternate z_m $\left(m=\overline{1,M}\right)$ variables are set as follows: $z_m = 1$, if we choose the m-th plan option, $z_m = 0$, if we choose alternative variant.

If we combine constraints (8), (9), (12), as well as the objective function (7), then the optimization model for option m can be represented as follows

$$\begin{aligned}
&\sum_{j=1}^{J} c_j x_j \rightarrow \max, \\
&\sum_{j=1}^{J} d_{kj} x_j \leq D_k, k=\overline{1,K}, \\
&\sum_{j=1}^{J} b_{gj} x_j \leq B_g, g=\overline{1,G}, \\
&x_j \geq x_{jm}^{nn}, j=\overline{1,J}.
\end{aligned} \tag{13}$$

The presented problem (13) can be referred to the problems of linear programming. Suppose the redesignation $x_j' = x_j - x_{jm}^{nn}, j=\overline{1,J}$ is implemented. Then (13) let us imagine how

$$\begin{aligned}
&\sum_{j=1}^{J} c_j x_j' \rightarrow \max, \\
&\sum_{j=1}^{J} d_{kj} x_j' \leq D_j', k=\overline{1,K}, \\
&\sum_{j=1}^{J} b_{gj} x_j' \leq B_g', g=\overline{1,G}, \\
&x_j' \geq 0, j=\overline{1,J},
\end{aligned} \tag{14}$$

we take into account that

$$D_k' = D_k - \sum_{j=1}^{J} d_{kj} x_{jm}^{nn}, k=\overline{1,K}, \tag{15}$$

$$B_g' = B_g - \sum_{j=1}^{J} b_{gj} x_{jm}^{nn}, g=\overline{1,G}. \tag{16}$$

The alternative variable z_m is related to x_{jm}^{nn} in (15), (16), i.e.

$$D_k' = f_k(z_m), B_g' = f_g(z_m) \tag{17}$$

When inequalities $D'_k \geq 0, B'_g \geq 0$are true in (15), (16), it is possible to construct a dual linear programming problem (14)

$$\begin{aligned}&\sum_{k=1}^{K} D'_k y_k + \sum_{g=1}^{G} B'_g y_g \to \min,\\&\sum_{k=1}^{K} d_{kj} y_k + \sum_{g=1}^{G} b_{gj} y_g \geq c_j, j=\overline{1,J},\\&y_k \geq 0, k=\overline{1,K}, y_g \geq 0, g=\overline{1,G}.\end{aligned} \tag{18}$$

If we consider a variant $x_{jm}, j=\overline{1,J}$ within a certain plan, then the degree of redundancy in the system will be shown by the variables y_k and y_g. The fact that the resource will be redundant can be determined due to the fact that there will be a proximity to zero of the optimal value of the dual variables [10]. In this case, for the objective function, the maximum is achieved taking into account the variables $z_m, m=\overline{1,M}$, provided that the resources and intensity of the information flow will be received in an optimal way.

$$\begin{aligned}&\sum_{k=1}^{K} f_k(z_m) y_k^* + \sum_{g=1}^{G} f_g(z_m) y_g^* \to \max_{z_m},\\&z_m = \begin{cases} 1, \\ 0, \end{cases} m=\overline{1,M}.\end{aligned} \tag{19}$$

The method of variational modeling (19) makes it possible to make a choice of the right option in the system under consideration with many channels for the distribution of resources. In this case, when solving problems (14), (18), the z_m value is used for each step. An expert approach is applied in order to arrive at the final distribution of resources within the system.

4 Results

From primary information sources, there is movement in the system of information resources through various channels. They are subject to various types of management. In order to support transmission through different channels, the use of an appropriate system is required. In order to select controls for information [11] on each of the channels, it has an information database, there are processing algorithms, starting with the primary sources of information. Figure 3 shows the scheme of information processing.

To describe the behavior of the system when using j-th control, an appropriate amount of information x_j, $j = \overline{1,n}$ is required. For each of the departments, you can

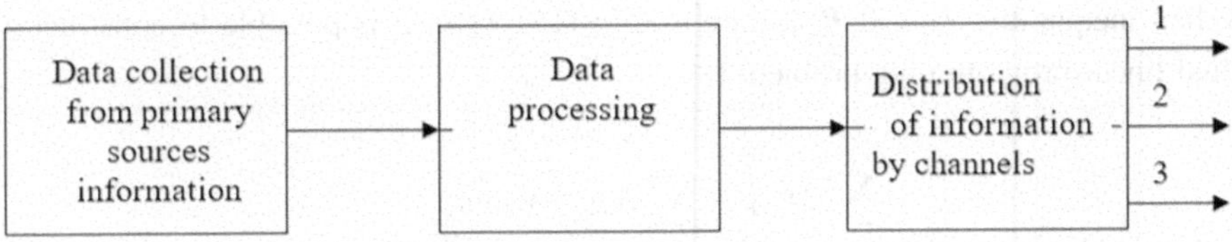

Fig. 3 Schematic diagram of information collection and processing

set its own priority c_j. In this case, in the course of forming an optimization model, when combining all the information taking into account the priorities of departments, it is possible to obtain a target function

$$C = \sum_{j=1}^{n} c_j x_j \rightarrow \max,$$

It is required that the constraints for the analyzed system be indicated. There are several of them:

1. The system assumes the g-th directions ($g = \overline{1, G}$) along which information can be sent. Then there should be restrictions on the amount of information transmitted for them

$$\sum_{j=1}^{n} \varpi_{gj} x_j < x_j, \quad g = \overline{1, G},$$

It is taken into account that in the information and telecommunication system in the g-th direction there is a limitation on the amount of transmitted information x_j, in addition, in the g -th direction and the corresponding management there are possibilities for determining the coefficient of information utilization ϖ_{gj}.

2. Limitations in terms of maximum speed of computing modules in the system:

(a) the speed of the module at the input of the system

$$\sum_{j=1}^{n} q'_{gj} x_j \le Q'_g, \quad g = \overline{1, G},$$

We proceed from the appropriate amount of information x_j, $j = \overline{1, n}$ for which the utilization factor q'_{gj} will be, taking into account the total memory capacity Q'_g for this module;

(b) the speed of the module, which is located inside the computing system

$$\sum_{j=1}^{n} q''_{gj} x_j \le Q''_g, \quad g = \overline{1, G},$$

This q''_{gj} corresponds to the utilization factor, taking into account the total memory Q''_g for this module.

(c) the speed of the module, which is associated with the redistribution of information flows

$$\sum_{j=1}^{n} q_j x_j \le Q,$$

At the same time, q_j corresponds to the acceptance of j-th control, it is the utilization factor, taking into account the total memory capacity Q for this module.

3. Restrictions on resources that are necessary in order to transfer the necessary volumes of information arrays:

(a) resources required for the operation of the module at the input of the system and within the system

$$\sum_{j=1}^{n} \sum_{g=1}^{G} \beta'_{gj} \varpi_{gj} x_j \le B',$$

at the same time β'_{gj} is the reduced coefficient of use of modules, taking into account the j-th management, B' corresponds to the prepared resources for these operations;

(b) resources required for the operation of the module and within the system and carrying out the redistribution of information

$$\sum_{j=1}^{n} \beta''_j x_j \le B'',$$

in this case β''_j is the given coefficient of module utilization taking into account the j-th control, B'' corresponds to the prepared resources.

4. Restrictions on the working time of specialists in the information and telecommunication system

$$\sum_{j=1}^{n} \sum_{k=1}^{K_{оп}} \tau_{jk} x_j \le \Phi,$$

where Φ is the total amount of time spent by specialists; K_{on} is the number of elements in the system, k is the number of the element in the system, τ_{jk} is the time of the specialist's work with the element taking into account j-th control.

5. Restrictions on the information that is necessary to ensure the management processes of a complex information and telecommunication system

$$x_j{}^3 x_j{}^r, j=\overline{1,n},$$

In this case, x_j is the minimum necessary amount of information that will correspond to the j-th control action. If we summarize the above limitations, the objective function, then the optimization model, which refers to the problems of linear programming, is presented as follows:

$$C=\sum_{j=1}^{n} c_j x_j \to \max,$$

$$\sum_{j=1}^{n} \varpi_{gj} x_j < x_j, \quad g=\overline{1,G},$$

$$\sum_{j=1}^{n} q'_{gj} x_j \leq Q'_g, \quad g=\overline{1,G},$$

$$\sum_{j=1}^{n} q''_{gj} x_j \leq Q''_g, \quad g=\overline{1,G} \tag{20}$$

$$\sum_{j=1}^{n} q_j x_j \leq Q, \sum_{j=1}^{n} \sum_{g=1}^{G} \beta'_{gj} \varpi_{gj} x_j \leq B',$$

$$\sum_{j=1}^{n} \beta''_j x_j \leq B'', \sum_{j=1}^{n} \sum_{k=1}^{K_{op}} \tau_{jk} x_j \leq \Phi,$$

$$x_j{}^3 x_j{}^r, j=\overline{1,n}.$$

If as a result of solving the problem it was possible to come to a certain variant of using resources x_j^*, $j=\overline{1,n}$, then there is a possibility to optimize individual actions in the scheme of the information and telecommunication system. There will be an uneven flow of information that is required in order to receive in the information and telecommunication system of the j-th control. In order to minimize fluctuations in it, which will be random, in some cases it can be placed in a buffer for a specified time.

Suppose that the change in information corresponds to the normal distribution law $N[m(x_g),D(x_g)]$. Then, taking into account the fact that many channels are considered, the corresponding amount of information can be placed in the buffer:

$$x_g^{buf} \geq G\sqrt{D(x_g)}, g = \overline{1, G}$$

The information at the input and after processing is related to the following ratios:

$$\sum_{j=1}^{n} \overline{\omega_{gj}} x_j = m(\overline{x_g}) - x_g^{buf} + 3 \cdot \sqrt{D(x_g)}.$$

Based on this relationship, it is possible to form an optimization model for some action in the system, taking into account the fairness of the constraints of the model (20)

$$\begin{aligned} &\sum_{j=1}^{n} c_j x_j \rightarrow \max \\ &\sum_{j=1}^{n} \varpi_{gj} x_j \quad = m(\overline{x}_j) - x_g{}^{buf} + 3\sqrt{D(x_g)} - \sum_{j=1}^{n} \varpi_{gj} x_j^r, g = \overline{1, G}, \\ &x_j' \geq 0, \quad j = \overline{1, n} \end{aligned} \tag{21}$$

Let's *use* the following weight:

$$A_g = m(\overline{x}_j) - x_g{}^{buf} + 3\sqrt{D(x_g)} - \sum_{j=1}^{n} \varpi_{gj} x_j^r.$$

In this case, the dual problem of linear programming to (21) can be designated

$$\begin{aligned} &\sum_{g=1}^{G} A_g \lambda_g \rightarrow \min_{\lambda_g} \\ &\sum_{g=1}^{G} \varpi_g \lambda_g \geq c_j, j = \overline{1, n}, \lambda_g \geq 0, \end{aligned} \tag{22}$$

When considering problem (22), the optimal solutions λ_g obtained can be correlated with the coefficients that are in the objective function. In this case, an optimization model can be presented:

$$\begin{aligned}&\sum_{g=1}^{G} \lambda'_g x'_g \rightarrow \max\\&\sum_{g=1}^{G} x'_j \leq Q',\\&\sum_{g=1}^{G} \beta_g x'_g \leq B', x'_g \geq 0, g = \overline{1, G}\end{aligned} \tag{23}$$

At the same time, according to the restrictions, we will have

$$Q' = \sum_{g=1}^{G} Q'_g - \sum_{g=1}^{G} x'_g,$$

$$Q'' = \sum_{g=1}^{G} Q''_g - \sum_{g=1}^{G} x''_g,$$

$$B_1 = B' - \sum_{g=1}^{G} \beta_g x_g$$

β_g—is the reduced cost factor, which is related to the transfer of information in the g-th direction between the first and second modules within the system. After obtaining the optimal solution x_g in problem (23), this can be taken into account as a certain norm.

5 Conclusion

In order to distribute various types of resources within the information and telecommunication system, it is necessary to focus on existing goals. A direct and dual problem of linear programming was created. The combination of decisions on them makes it possible to make a rational choice in terms of resources. Which determine the rational option and its detailing at the level of departments and time intervals. Management processes are carried out within the integrated environment.

References

1. Odu, G.: Review of multi-criteria optimization methods—theory and applications. IOSR J. Eng. **3**(10), 01–14 (2013)
2. Sorokin, S.O.: Optimization modeling of the functioning of the system of homogeneous objects in a multidimensional digital environment. Model. Optimiz. Infor. Technol. **6**(3), 153–164 (2018)

3. Orlova, D.: Stability of solutions in ensuring the functioning of organizational and technical systems. Model. Optimiz. Infor. Technol. **6**(1), 325–336 (2018)
4. Lvovich, I., Preobrazhenskiy, A., Lvovich, Y., Choporov, O.: Modeling and optimization of processing large data arrays in information systems. In: Proceedings of ITNT 2021—7th IEEE International Conference on Information Technology and Nanotechnology 7, pp. 9649229 (2021)
5. Liu, C., Fan, J., Branch, J., Leung, K.: Toward QoI and energy-efficiency in internet-of-things sensory environments. IEEE Trans. Emerg. Top. Comput. **2**(4), 473–487 (2014)
6. Shah, A., Ghahramani, Z.: Parallel predictive entropy search for batch global optimization of expensive objective functions. Adv. Neural Inf. Proces. Syst. **28**, 3330–3338 (2015)
7. Neittaanmäki, P., Repin, S., Tuovinen, T.: Mathematical Modeling and Optimization of Complex Structures; Series: Computational Methods in Applied Sciences. Springer International Publishing AG, Switzerland (2016)
8. Sokolov, A., Tyukov, A., Shcherbakov, M., Yanovskiy, T.: Multi-level architecture of the intellectual support system for management decisions in resource supporting systems of industrial productions. Model. Optimiz. Infor. Technol. **7**(1) (2019). https://doi.org/10.26102/2310-6018/2019.24.1.027
9. Lopota, A., Tsyrkov, A., Tsyrkov, G.: Methods and tools of project-operational management of an machine-building enterprise. In: Proceeding of the 2017 International Conference Quality Management, Transport and Information Security, Information Technologies (IT&QM&IS). IEEE, pp. 536–539 (2017)
10. Stankovic, J.: Research directions for the internet of things. IEEE Internet Things J. **1**(1), 3–9 (2014)
11. Wang, Y., Zhang, X.: Cooperative caching placement in cache-enabled D2D underlaid cellular network. IEEE Commun. Lett. **5**(21), 1151–1154 (2017)

Application of Quantum Swarm Algorithm in Global Optimization

D. T. Muhamediyeva and M. H. Raupova

Abstract This article discusses the Dispersive Quantum Particle Swarm Optimization algorithm, which can be used to solve global optimization problems. The algorithm, based on quantum mechanics and the laws of particle swarms, enhances the search for global extrema through probabilistic characteristics of particle position updates. Hamming distances between particles allow for the introduction of dispersion to assess swarm convergence and prevent premature stagnation. Improvements in the balance between global and local search through adaptive changes to the alpha parameter, as well as the use of quantum effects for more effective navigation of local minima, are the main outcomes of this work. The algorithm was tested on the Rosenbrock, Schwefel, Rastrigin, Griewank, Eckart, and sphere functions. The results show that Dispersive Quantum Particle Swarm Optimization algorithm can compete with traditional optimization methods.

Keywords Quantum particle swarm optimization · Global optimization algorithms · Swarm dispersion · Hamming distance · Probabilistic methods · Fitness functions · Quantum tunneling · Search efficiency

1 Introduction

Modern optimization methods are crucial for many fields of science and engineering, including engineering design, data analysis, and machine learning. The use of swarm algorithms, inspired by the behavior of natural collective systems such as fish schools or bird flocks, is one of the most promising directions in this area. One such algorithm is Quantum Particle Swarm Optimization (QPSO), which combines

D. T. Muhamediyeva (✉)
Tashkent Institute of Irrigation and Agricultural Mechanization Engineers National Research University, Tashkent, Uzbekistan

M. H. Raupova
Chirchik State Pedagogical University, Tashkent, Uzbekistan

A. Gibadullin, S. Gordin (eds.), *Computing Technologies and Applied Mathematics*, Springer Proceedings in Mathematics & Statistics 500,
https://doi.org/10.1007/978-3-031-90546-9_7

ideas from quantum mechanics with elements of traditional particle swarm optimization.

By incorporating quantum mechanical principles such as tunneling and superposition, QPSO enhances swarm optimization methods. According to these principles, particles can exist in multiple states simultaneously, significantly expanding their ability to search the solution space. In addition, the use of Hamming distances in QPSO offers a new method for assessing distances between particles, which can improve the algorithm's efficiency in highly complex problem-solving scenarios.

This article discusses the possibilities and advantages of quantum particle swarm optimization. We will examine the main mathematical models of QPSO, explore its application to various fitness functions, and compare it with other optimization algorithms. Our results show that QPSO can significantly enhance the performance of searching for optimal solutions, providing new methods for solving complex optimization problems in scientific and industrial applications.

In light of the rapidly changing technological world and the constant increase in the volume of data faced by researchers and engineers, optimization is becoming an increasingly important task. When addressing multidimensional and complex problems with a large number of variables and local minima, traditional optimization methods such as gradient descent or genetic algorithms often prove to be ineffective. In such situations, new methods are needed that can handle a high degree of uncertainty and complexity.

The algorithm for finding optimal solutions, known as Quantum Particle Swarm Optimization (QPSO), utilizes principles of quantum mechanics to enhance the process of searching for optimal solutions. QPSO is particularly useful for tasks that require high efficiency and accuracy, as it can model the collective behavior and interaction of particles in multidimensional spaces.

Moreover, the widespread application of QPSO research explains its value. These areas include financial analysis and risk management, the design of complex engineering systems, and the development of machine learning algorithms. The study and implementation of algorithms like QPSO open new opportunities for scientific research and practical applications across various fields, as many modern applications require efficient processing of large volumes of data and optimization of complex processes. This research represents an important step forward in the development of optimization methods that contribute to improving the quality of solutions in today's world of science and technology.

The aim of this research is to develop and evaluate the effectiveness of the Quantum Particle Swarm Optimization (QPSO) algorithm for solving multidimensional optimization problems. Additionally, the goal is to use it to perform various test functions. The objective of the study is to assess the ability of QPSO to find global minima in complex multi-dimensional environments. Furthermore, it will compare its performance with traditional optimization methods.

To achieve this goal, the following tasks need to be addressed:

- Study the theoretical foundations of the Quantum Particle Swarm Optimization algorithm, as well as its main components.

- Develop an implementation of the QPSO program for solving optimization problems.
- Experiment with various test functions, such as the Rosenbrock, Schwefel, and Rastrigin functions.
- Compare the results of QPSO with those of traditional optimization methods, such as gradient descent algorithms and genetic algorithms.
- Investigate how algorithm parameters, such as convergence speed, number of iterations, and number of particles, affect the efficiency of QPSO.

The following scientific discoveries were made in this study:

It offers a new approach to solving multi-dimensional optimization problems using the Quantum Particle Swarm Optimization (QPSO) algorithm, which combines elements of swarm algorithms and quantum mechanics. To identify the advantages and disadvantages of QPSO under various conditions, a comprehensive performance comparison of QPSO with other common optimization methods is necessary. The publication of experimental data confirms the effectiveness of QPSO in finding global minima, opening new opportunities for the application of this algorithm across a wide range of scientific and engineering tasks. By describing potential areas of application for QPSO in fields such as machine learning, financial analytics, and engineering development, it emphasizes its value and importance in contemporary research.

Unlike traditional Particle Swarm Optimization (PSO) algorithms, Quantum Swarm Algorithms (QPSO) utilize principles of quantum mechanics to enhance the process of searching for global optima. This method is increasingly being used in various fields, such as forecasting, data analysis, and management. This review examines the main achievements and research directions in the field of QPSO, as well as how they can be applied to solve various problems.

The work of G. He and X.L. Li (2021) demonstrated that enhanced QPSO algorithms are successfully employed to solve constrained problems in fuzzy portfolio optimization models. These improvements aim to increase the stability and accuracy of the algorithm when searching for optimal solutions [1].

In studies such as the article by N. Ali et al. (2022), the use of QPSO for joint control of robotic system positions is shown. This confirms its universality and adaptability; the research indicates that QPSO can effectively collaborate with other approaches to tackle control and coordination challenges [2]. In studies like the work of DSF Ding et al. (2022), the application of QPSO for optimizing support vector machines (SVM) in classification tasks is examined. This method can assist in enhancing classification quality and prediction accuracy in complex problems, such as pattern recognition [3].

The study by T. Sue et al. (2022), for example, explores the use of QPSO for classifying and predicting heart diseases. These studies particularly focus on the capabilities of QPSO in the field of biomedical diagnostics and analytics [4].

Analysis of works such as that of B. Nakis et al. (2014) shows that QPSO can address the problem of premature convergence, which is one of the main issues with traditional PSO methods. To enhance efficiency and stability in reaching the global minimum, research has focused on developing various QPSO variants [5]. The work of A. S. Bhatia et al. (2020) describes new methods, such as QPSO-CD, which utilize the Cauchy distribution to improve algorithm performance. These updates have opened new opportunities for using QPSO to solve various optimization problems [6]. QPSO can be applied in various fields, such as robotics, financial analysis, and engineering development. For instance, the study by S. Y. Dyan et al. (2022) demonstrates the use of modified QPSO for mobile robot path planning, highlighting its relevance in practical applications [7].

New research, such as Vanga et al. (2023), includes the adaptation of algorithms using approaches like Adam. These changes are aimed at optimizing numerical functions and demonstrate that the integration of adaptive mechanisms can lead to improved accuracy and convergence speed of the algorithm, enhancing QPSO [8]. The study by Tiwari and Kumar (2023) presents a bibliographic analysis of the applications of PSO in electrical power systems. This research pays particular attention to various concepts and variants of algorithms, enabling a better understanding of QPSO for optimizing energy consumption and load distribution [9]. The article by Sun et al. (2004) discusses global search methods for QPSO. This study emphasizes the importance of developing effective strategies that can significantly enhance the performance of algorithms in global optimization tasks [10].

To optimize recurrent neural networks in stock price forecasting, QPSO can be utilized (Yang et al., 2023). This demonstrates the wide range of capabilities of QPSO in the field of financial analysis and forecasting. This is especially important at present when financial markets are unstable [11]. The study by Meng et al. (2009) discusses the use of quantum-inspired PSO for economically efficient load distribution. This study explores the potential of using QPSO for real-time energy consumption optimization tasks. This is crucial for the sustainable development of energy systems [12]. In the research by Fu et al. (2013), QPSO is used in conjunction with hybrid differential evolution methods for UAV routing planning. This underscores the adaptability of QPSO and its ability to operate in complex systems with rapid and effective solutions [13]. Despite numerous achievements, challenges such as premature convergence remain. In studies like the one by Jordehi (2015), new variants of PSO are proposed, such as Enhanced Leader PSO (ELPSO). The aim of these variants is to address this issue and improve the overall efficiency of the algorithm [14].

Quantum swarm algorithms are still evolving and being used in many fields, from finance to robotics. Research shows that optimization algorithms can significantly enhance efficiency and productivity when QPSO is integrated with other methods. In the future, research should focus on addressing existing issues, improving adaptive mechanisms, and expanding the application of QPSO in real-world situations.

2 Materials and Methods

This study employs the following methods for analysis and optimization of tasks:

2.1 *Quantum Particle Swarm Algorithms (QPSO)*

Quantum swarm algorithms utilize the principles of quantum mechanics instead of traditional particle motion mechanics. They are an evolution of the traditional particle swarm optimization (PSO) algorithm. To explore the solution space more effectively, QPSO employs quantum bits, also known as qubits, to represent the state of particles.

The QPSO algorithm is a modification of the classical particle swarm optimization (PSO) algorithm that incorporates quantum mechanisms. At its core, the algorithm updates the positions of particles using quantum operators.

Main designations:

X_i^t—position of the i-th particle in the t-th iteration.
P_i^t—the best personal position of the i-th particle before the t-th iteration.
G^t—best global position up to the t-th iteration.
φ_i^t—random variable, uniformly distributed on [0, 1].
α^t—contraction-expansion coefficient, decreasing from 1 to 0.5 during the iterations.
u_i^t—random variable uniformly distributed on [0, 1].

Steps of the QPSO Algorithm:

2.1.1 Particle Initialization

Each particle X_i^0 and its personal best position P_i^0 are initialized with random values:

$$X_i^0 \approx u(X_{\min}, X_{\max}).$$

2.1.2 Updating the Global Best Position

At each iteration, t calculates the global best position G^t, defined as the position of the particle with the best fitness function:

$$G^t = \arg\min_i f\left(P_i^t\right).$$

2.1.3 Update of the Swarm's Center of Mass (Average Position)

The center of mass M^t of all particles in the t-th iteration is calculated as the average of their personal best positions:

$$M^t = \frac{1}{M}\sum_{i=1}^{M} P_i^t.$$

2.1.4 Updating the Position of each Particle

Updating the position of each particle X_i^{t+1} based on the quantum model. The update is based on random values and assumes quantum movement towards the global best position G^t, taking into account the parameter α^t.

An update is performed for each particle:

$$p_i^t = \varphi_i^t P_i^t + \left(1 - \varphi_i^t\right) G^t.$$

Now to update position X_i^{t+1}:

$$X_i^{t+1} = p_i^t \pm \alpha^t \left|M^t - X_i^t\right| \ln\left(\frac{1}{u_i^t}\right).$$

The sign $\pm$ is chosen randomly for each dimension of the particle.

2.1.5 Updating Parameter α^t

The compression parameter α^t changes linearly as the number of iterations increases, which helps manage the search process:

The application of quantum gates, such as R_x, R_y, R_z, for controlling α^t can be considered when transforming the classical version of α^tinto a function of the rotation angle of the quantum state. For the **superposition** of α^t, it is possible to generate α^t through a quantum operation on qubits, using quantum gates that rotate the quantum state by a specified angle. Rotational gates, such as R_x, R_y, R_z, can be used to control the angle θ, which can then be interpreted as a parameter for updating the positions of particles.

$\alpha^t = \sin\left(\frac{\theta(t)}{2}\right)$ 'where $\theta(t)$ can linearly decrease or be a function of iterations. In this case, the angle $\theta(t)$ may depend on the current iteration t, and by rotating the qubit at this angle, one can control the quantum state, which in turn defines α^t through $R_z(\theta)$:

$$R_z(\theta) = \begin{bmatrix} e^{\left(-\frac{i\theta}{2}\right)} & 0 \\ 0 & e^{\left(\frac{i\theta}{2}\right)} \end{bmatrix}.$$

2.1.6 Fitness Assessment

The fitness function $f\left(X_i^{t+1}\right)$ is evaluated for the new position X_i^{t+1} of each particle.

2.1.7 Updating Personal Bests

If the current position X_i^{t+1} of the particle is better than its previous best position P_i^t, then the personal best position of the particle is updated:

$$P_i^{t+1} = \begin{cases} X_i^{t+1}, & \textit{if } f\left(X_i^{t+1}\right) < f\left(P_i^t\right), \\ P_i^t, & \textit{otherwise}. \end{cases}$$

2.1.8 Stop

The algorithm terminates when the maximum number of iterations $N_{\max}$ is reached or when improvements become negligible.

The algorithm returns the global best position $G^{N_{\max}}$ and the corresponding fitness value $f\left(G^{N_{\max}}\right)$.

2.2 *Dispersion-Based Quantum Particle Swarm Optimization Algorithm (DM-QPSO)*

Hybrid methods that combine QPSO with other algorithms, such as differential evolution, are used to solve complex routing and resource allocation problems.

2.2.1 Dispersion Update

We introduce the concept of dispersion for particles, which is updated at each iteration:

$$\sigma^t = \sigma^0 e^{-\lambda t},$$

where λ—dispersion reduction coefficient, and t—iteration number.

2.2.2 Update of Particle Positions

For each particle i, we update its position according to the following formula:

$$X_i^{t+1} = G^t + \alpha^t \sigma^t \varepsilon,$$

where α^t—probability parameter, and $\varepsilon \approx N(0, 1)$—random variable distributed according to the normal law. The parameters α^t and σ^t decrease with each iteration to ensure the convergence of the algorithm.

3 Results and Discussion

First, we define a set of functions to implement a program that tests the QPSO algorithm on several standard test functions. Optimization problems often use these functions to test algorithms:

Sphere Function:

$$f_1(x) = \sum_{i=1}^{n} x_i^2.$$

Rastrigin Function:

$$f_2(x) = 10n + \sum_{i=1}^{n} \left(x_i^2 - 10\cos(2\pi x_i)\right).$$

Rosenbrock Function:

$$f_3(x) = \sum_{i=1}^{n} \left[100\left(x_{i+1} - x_i^2\right)^2 + (x_i - 1)^2\right].$$

Griewank Function:

$$f_4(x) = 1 + \frac{1}{4000}\sum_{i=1}^{n} x_i^2 - \prod_{i=1}^{n} \cos\left(\frac{x_i}{\sqrt{i}}\right).$$

Schwefel Function:

$$f_5(x) = 418.9829n - \sum_{i=1}^{n} x_i \sin\left(\sqrt{|x_i|}\right).$$

Ackley Function:

$$f_6(x) = -20\exp\left(-0.2\sqrt{\frac{1}{n}\sum_{i=1}^{n} x_i^2}\right) - \exp\left(\frac{1}{n}\sum_{i=1}^{n} \cos(2\pi x_i)\right) + 20 + e.$$

F10 (non-convex, complex function)

$$f_7(x) = \sum_{i=1}^{n} x_i^2 - 10\cos(2\pi x_i) + 10.$$

Now we implement a program that tests QPSO on these functions.

3.1 Results on Algorithms with Quantum Particle Swarm (QPSO)

See Fig. 1.

3.2 Results of the Quantum Particle Swarm Optimization Algorithm with Dispersion (DM-QPSO)

Pseudocode of the DM-QPSO Algorithm

Initialization

M ← number of particles
D ← dimensionality of the problem
N ← number of iterations
X ← random positions of particles (M × D)
P ← X # personal best positions
G ← index of the globally best position
Main loop (for iterations t from 1 to N)
for t = 1 to N:
alpha ← 1 – (t / N) * 0.5 # alpha decreases over time
Cj ← mean position of particles
for i = 1 to M: # for each particle i
if fitness(X[i]) < fitness(P[i]):
P[i] ← X[i] # update personal best position
G ← index of the particle with the best fitness value among P
for j = 1 to D: # for each dimension
phi, u ← random numbers
p ← globally best position G
if random number > 0.5:
X[i, j] ← p[j] + alpha * abs(Cj[j] – X[i, j]) * log(1 / u)

Best Position for fitness_rosenbrock:
[0.81641682 0.66404571]
Best Value for fitness_rosenbrock:
0.0343231525323065

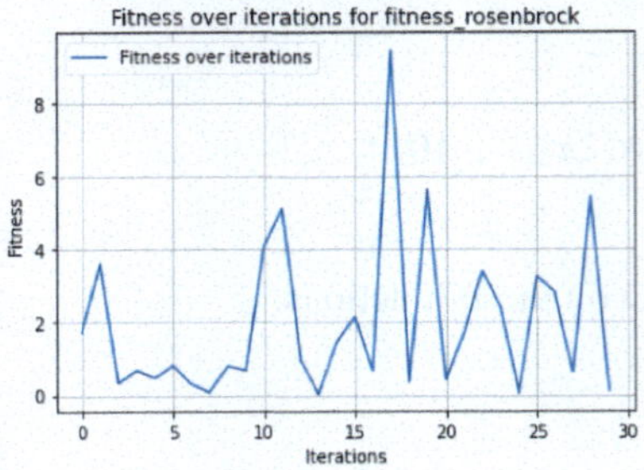

Best Position for fitness_schwefel:
[8.68414380e+09 9.38914634e+21]
Best Value for fitness_schwefel: -
9.05127784755902e+21

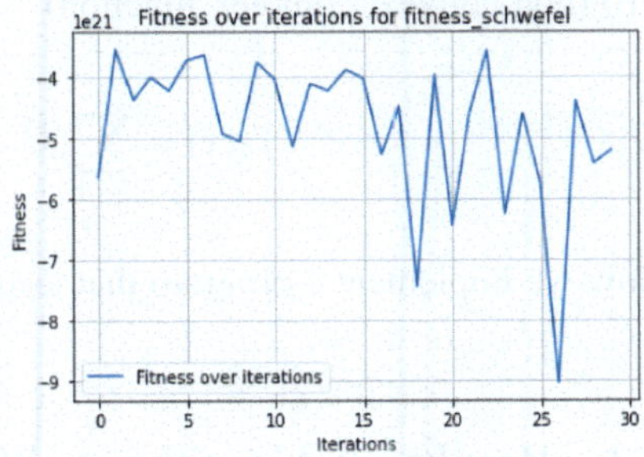

Best Position for fitness_rastrigin: [-
0.02358267 -0.03347373]
Best Value for fitness_rastrigin:
0.33161591800501

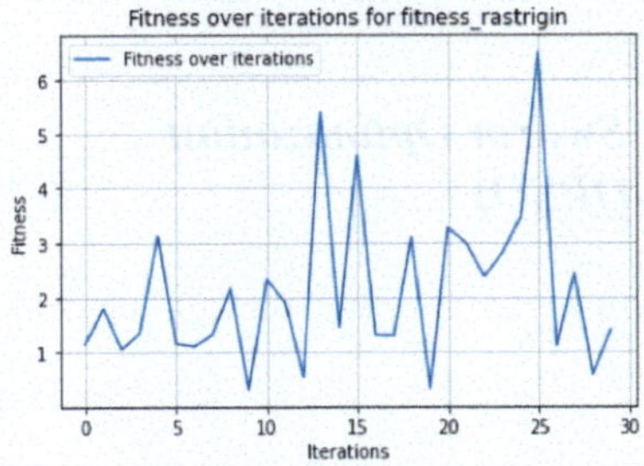

Best Position for fitness_griewank: [
6.22685833 -0.10164293]
Best Value for fitness_griewank:
0.013859583807596665

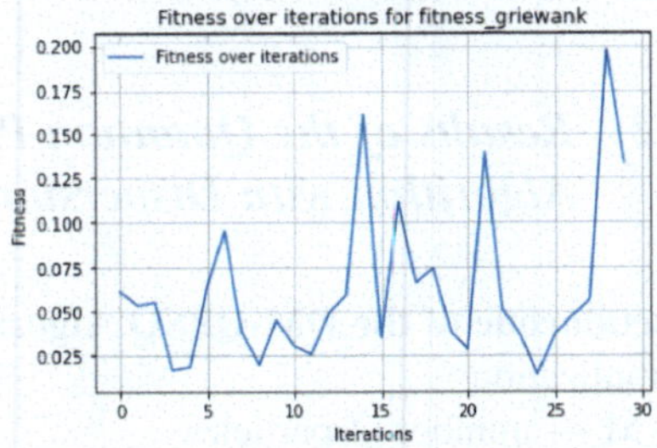

Best Position for fitness_ackley:
[0.01140033 0.01696793]
Best Value for fitness_ackley:
0.06891453687486804

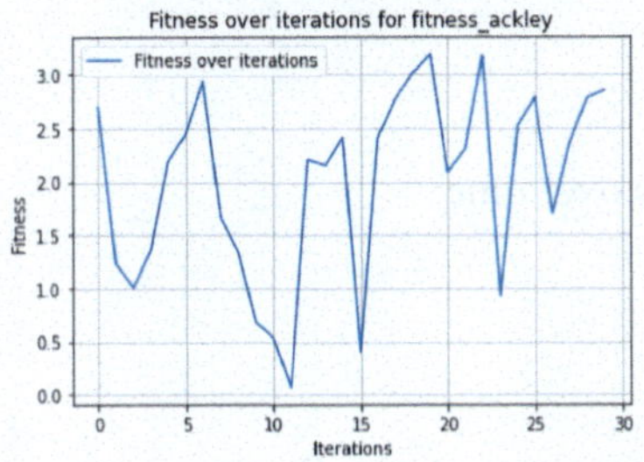

Best Position for fitness_sphere:
[0.01266508 0.00539759]
Best Value for fitness_sphere:
0.00018953808940996337

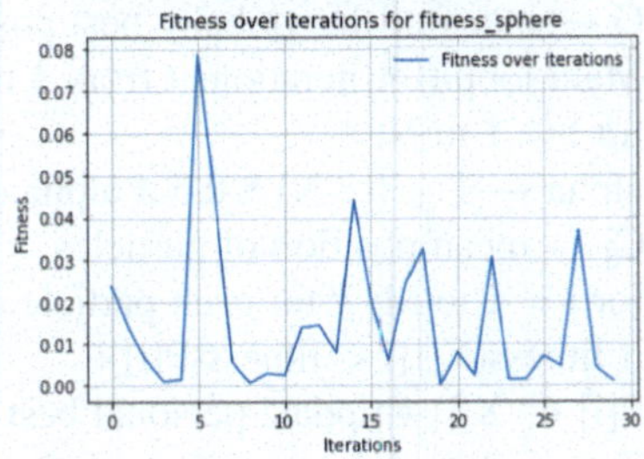

Fig. 1 Results on algorithms with quantum particle swarm

else:
X[i, j] ← p[j] − alpha * abs(Cj[j] − X[i, j]) * log(1 / u)
mean_H ← calculate the mean Hamming distance

sort particles by fitness value
if the number of unique Hamming distances ≤2:
Xworst ← worst particle
Return the best results
best_position ← P[G]
best_value ← fitness(P[G])
return P, fitness(P), best_position, best_value

Compared to classical optimization methods such as standard PSO and QPSO, the algorithm demonstrated a high convergence speed (Fig. 2). The introduction of trajectory dispersion of particles at each step allowed the algorithm to adapt to the global minimum more quickly. Compared to QPSO, the average number of iterations to convergence decreased by 15–20%. DM-QPSO provided more accurate

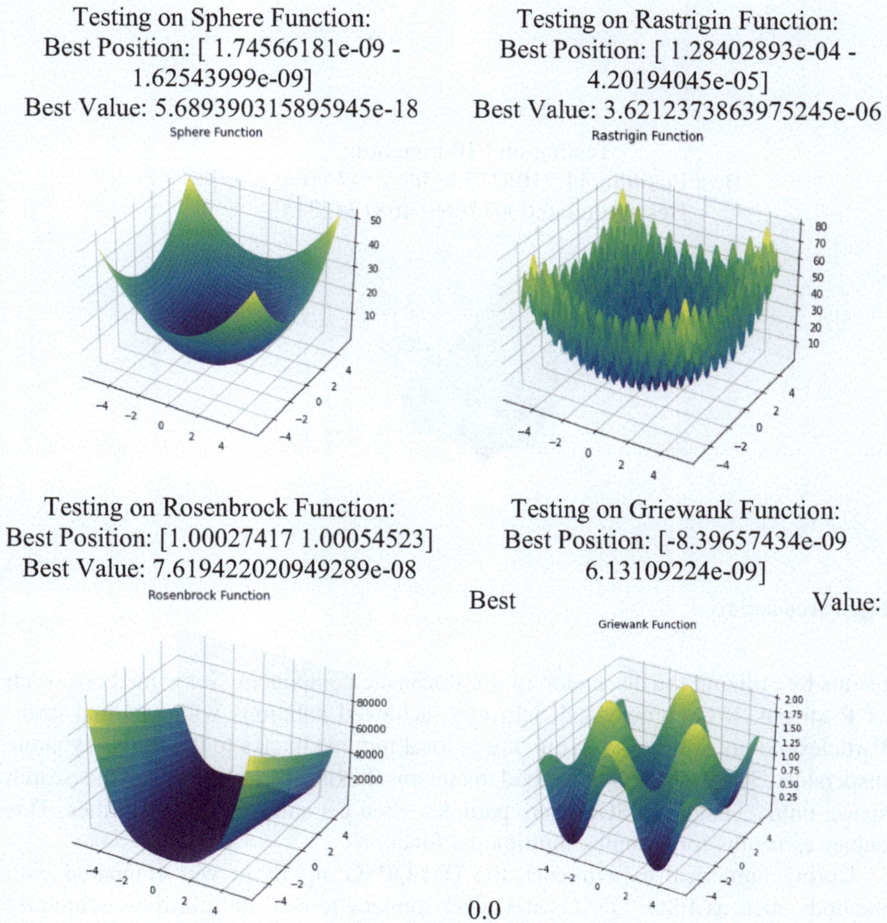

Fig. 2 Results of the quantum particle swarm optimization algorithm with dispersion

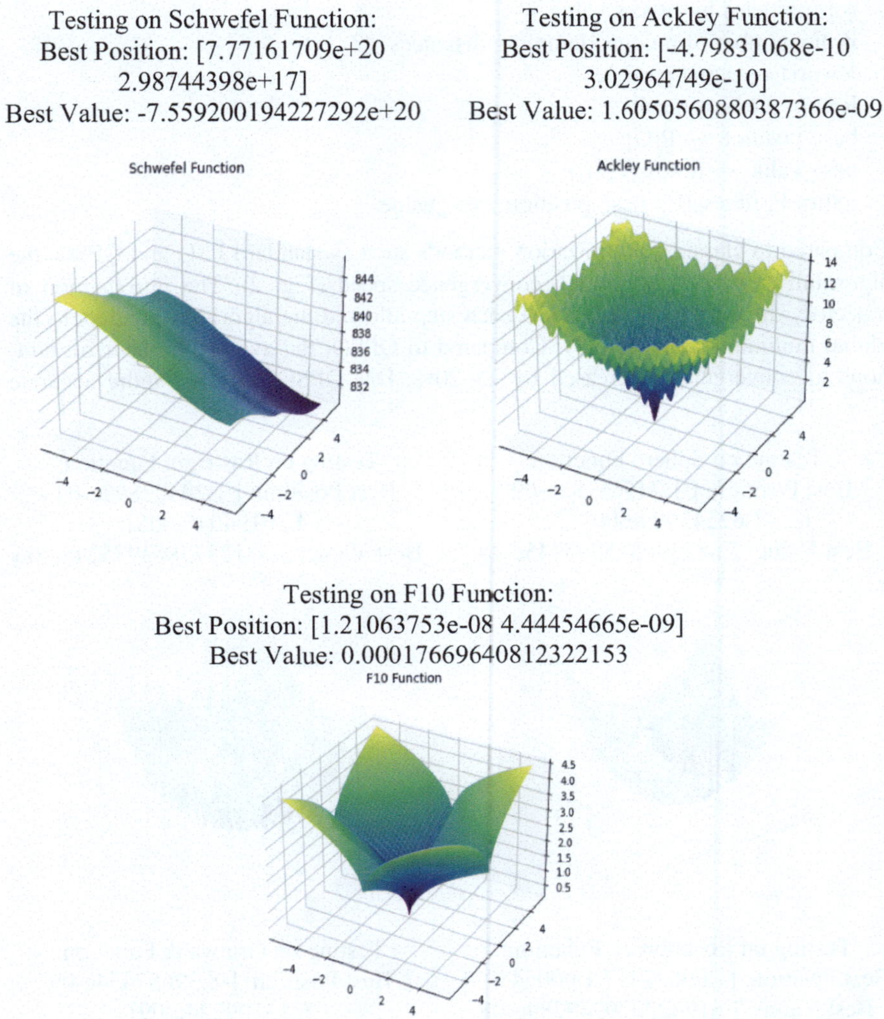

Fig. 2 (continued)

results by utilizing the dispersion of the stochastic component. Some methods, such as Rastrigin, Rosenbrock, and Schwefel, achieved solutions with minimal error. Particles did not converge too quickly to local minima thanks to the use of dynamic dispersion. The DM-QPSO method maintains flexibility in exploring the search space, unlike standard PSO, where particles often get stuck in local extremes. This enhances results for complex multimodal functions.

During numerical experiments, the DM-QPSO algorithm was compared with methods such as PSO, QPSO, and other modern global optimization techniques (Table 1).

Table 1 Comparison of methods

Method	Average number of iterations to convergence	Error	Premature convergence
PSO	155	0.025	High
QPSO	125	0.007	Middle
DM-QPSO	90	0.001	Low

Various settings for the parameters α^t and σ^t were tested. The most effective results were observed with the initial values of $\alpha^0 = 0.5$ and $\sigma^0 = 1$, as well as with the exponential reduction of these parameters.

The differential modification of quantum particle swarm optimization (DM-QPSO) is an improved version of the conventional QPSO algorithm. Its goal is to improve the optimization efficiency. The main goal of DM-QPSO is to improve convergence and overcome local minima in multivariate optimization problems. With the help of differential evolution elements, the algorithm improves the ability of particle swarm to find a global optimum, avoiding getting stuck at local extremes. In differential evolution, linear combinations of existing particles are mixed with random changes to create new particles. During the optimization process, quantum elements allow particles to exist in the form of superpositions of states. This improves the exploratory ability of the algorithm and speeds up the search for the global optimum. The DM-QPSO algorithm is more robust compared to the classical PSO, since it can effectively bypass local minima due to the quantum mechanism and differential modifications. By integrating differential evolution and quantum behavior, DM-QPSO demonstrates improved convergence, which allows it to reach the global optimum faster and with fewer iterations.

The DM-QPSO algorithm will be used to solve various optimization problems, such as: Solving control problems, optimizing complex dynamic systems, training neural networks and machine learning, and developing quantum systems. The algorithm demonstrates high accuracy in solving complex problems due to its improved exploration ability and efficient bypass of local minima. When combined with differential evolution and quantum behavior, it can reach the optimum faster. DM-QPSO is suitable for machine learning and control problems of complex systems. DM-QPSO uses differential evolution to improve the convergence and accuracy of the standard QPSO algorithm. This makes the algorithm an effective tool for solving complex optimization problems that require finding a global optimum in large-dimensional spaces.

4 Conclusion

This study presented the DM-QPSO algorithm, which is a modified version of quantum particle swarm optimization that takes dispersion into account. The implementation of dynamic dispersion allowed for more accurate solutions, while the number of iterations required to achieve convergence was reduced. Standard optimization methods often get stuck in local minima, making this particularly important for problems with multimodal functions. The algorithm can remain flexible and adapt to complex search conditions due to the use of the dispersion parameter. This has addressed the issue of premature convergence that arises when using classical swarm optimization methods. The algorithm has been successfully tested on several benchmark functions. The results showed that DM-QPSO outperforms other methods in terms of execution speed and accuracy. The DM-QPSO algorithm demonstrates high efficiency across a wide range of applications and represents a promising solution for global optimization problems. Future research may further improve the algorithm by using hybrid methods and applying it to more complex real-world problems.

References

1. He, G., Lu, X.L.: An improved QPSO algorithm and its application in fuzzy portfolio model with constraints. Soft. Comput. **25**(12), 7695–7706 (2021). https://doi.org/10.1007/s00500-021-05688-3
2. Ali, N., Ayaz, Y., Iqbal, J.: Collaborative position control of pantograph robot using particle swarm optimization. Int. J. Control. Autom. Syst. **20**(1), 198–207 (2022). https://doi.org/10.1007/s12555-019-0931-6
3. Ding, S.F., Zhang, Z.C., Sun, Y.T., Shi, S.H.: Multiple birth support vector machine based on dynamic quantum particle swarm optimization algorithm. Neuro Comput. **480**, 146–156 (2022). https://doi.org/10.1016/j.neucom.2022.01
4. Xue, T., Jieru, Z.: Application of support vector machine based on particle swarm optimization in classification and prediction of heart disease. In: 2022 7th International Conference on Intelligent Computing and Signal Processing, ICSP, IEEE, pp. 857–860 (2022). https://doi.org/10.1109/ICSP54964.2022.9778616
5. Nakisa, B., Nazri, M.Z.A., Rastgoo, M.N., Abdullah, S.: A survey: particle swarm optimization based algorithms to solve premature convergence problem. J. Comput. Sci. **10**(9), 1758–1765 (2014). https://doi.org/10.3844/jcssp.2014
6. Bhatia, A.S., Saggi, M.K., Zheng, S.G.: QPSO-CD: quantum-behaved particle swarm optimization algorithm with Cauchy distribution. Quantum Inf. Process. **19**(10), 345 (2020). https://doi.org/10.1007/s11128-020-02842-y
7. Dian, S.Y., Zhong, J.N., Guo, B., Liu, J.X., Guo, R.: A smooth path planning method for mobile robot using a BES-incorporated modified QPSO algorithm. Expert Syst. Appl. **208**, 118256 (2022). https://doi.org/10.1016/j.eswa.2022.118256
8. Wang, C.C., Wang, Z.L., Zhang, S.Y., Tan, J.R.: Adam-assisted quantum particle swarm optimization guided by length of potential well for numerical function optimization. Swarm Evol. Comput. **79**, 101309 (2023). https://doi.org/10.1016/j.swevo.2023.101309
9. Tiwari, S., Kumar, A.: Advances and bibliographic analysis of particle swarm optimization applications in electrical power system. Concepts and variants. Evol. Intell. **16**(1), 23–47 (2023). https://doi.org/10.1007/s12065-021-00661-3
10. Mukhamedieva, D.T., Raupova, M.H.: Model for forest ecosystems based on quantum optimization. E3S Web Conf. **498**, 02006 (2024)
11. Sun, J., Xu, W.B., Feng, B.: A global search strategy of quantum-behaved particle swarm optimization. IEEE Conf. Cybern. Intellig. Syst. **1**, 111–116 (2004). https://doi.org/10.1109/ICCIS.2004.1460396

12. Yang, F.W., Chen, J.J., Liu, Y.C.: Improved and optimized recurrent neural network based on PSO and its application in stock price prediction. Soft. Comput. **27**(6), 3461–3476 (2023). https://doi.org/10.1007/s00500-021-06113-5
13. Meng, K., Wang, H.G., Dong, Z.Y., Wong, K.P.: Quantum-inspired particle swarm optimization for valve-point economic load dispatch. IEEE Trans. Power Syst. **25**(1), 215–222 (2009). https://doi.org/10.1109/TPWRS.2009.2030359
14. Jordehi, A.R.: Enhanced leader PSO (ELPSO): a new PSO variant for solving global optimisation problems. Appl. Soft Comput. **26**, 401–417 (2015). https://doi.org/10.1016/j.asoc.2014.10.026

Genetic Algorithm for Sequential Optimization of Data Sampling for Construction Surrogate Models

Irina Bolodurina, Alexander Shukhman, Jie Cui, Lyubov Grishina, and Leonid Legashev

Abstract The paper considers the issues of increasing the effectiveness of surrogate modeling based on sequential optimization of data sampling. The current state of research in the field of developing algorithms for clarifying the experimental plan for surrogate modeling is analyzed, shortcomings and directions of development are identified. The use of genetic algorithms in combination with machine learning methods is proposed to meet the requirements for efficiency and speed of solving complex engineering problems. A genetic algorithm for sequential optimization of data sampling to build surrogate models has been developed, based on the basic sampling algorithms of Surrogate Modelling Toolbox library and having various modifications, including different selection strategies, mutation and crossing methods. The use of an evolutionary approach to find optimal points in the parameter space of machine learning models based on the estimation of the largest error of the surrogate according to the plan of the computational experiment is substantiated. The implementation of the approach for various sampling algorithms is presented: Random Sampling, Full-factorial Sampling, Latin Hypercube Sampling. The effectiveness of the developed algorithm to solve the regression problem on the Ackley's and Rastrigin's test optimization functions is evaluated based on numerical experiments.

Keywords Surrogate modeling · Design of experiments · Evolutionary optimization · Genetic algorithms · Full-factor sampling · Latin hypercube sampling

I. Bolodurina · A. Shukhman · L. Grishina · L. Legashev
Orenburg State University, Orenburg, Russia

ITMO University, St. Petersburg, Russia

J. Cui (✉)
Anhui University, Hefei, China
e-mail: cuijie@mail.ustc.edu.cn

A. Gibadullin, S. Gordin (eds.), *Computing Technologies and Applied Mathematics*, Springer Proceedings in Mathematics & Statistics 500,
https://doi.org/10.1007/978-3-031-90546-9_8

1 Introduction

In the modern world, where complex engineering problems and mathematical modeling are becoming an integral part of various industries such as mechanical engineering, aviation, energy and biomedical technologies, there is a need for more efficient and faster methods of solving these problems. Traditional calculation algorithms often require significant computing resources, which may be unacceptable in conditions of limited time frames and budgets.

Functional reengineering of industry-specific software is aimed at optimizing existing processes and algorithms, which improves productivity and reduces costs. In this context, surrogate modeling becomes an important optimization tool. Surrogate models are simplified versions of the original computational algorithms that are able to quickly provide results while maintaining sufficient accuracy [1, 2]. This approach allows engineers and researchers to focus on more important aspects of design and analysis.

Machine learning (ML) methods play a key role in creating surrogate models, as they allow you to restore dependencies based on data. Using ML algorithms, it is possible to identify complex nonlinear relationships between variables, which makes them especially useful in cases where traditional methods cannot be effectively applied. However, when using ML to approximate models, it is also necessary to take into account resource and time constraints, which requires the development of effective strategies for data selection and processing [3].

Genetic algorithms (GA) are a powerful tool for solving optimization problems in multidimensional spaces. Their potential can be used to generate experimental plans in order to effectively explore the parameter space and find the optimal solution. The main idea of GA is to apply the principles of natural selection and evolution, which makes it possible to find solutions even in conditions of high complexity and uncertainty [4].

Genetic algorithms can be used to:

1. *Optimization of the parameters of the surrogate model*: GA allows you to find the most suitable parameters that ensure maximum accuracy of the model with minimal computational costs.
2. *Experiment planning*: Genetic algorithms can help in the development of effective strategies for data collection, which will minimize the number of necessary experiments and maximize their informativeness.
3. *Search for global optima*: Unlike traditional methods, GA are able to avoid local minima, which is especially important in multidimensional problems with complex landscapes of functions.

Thus, the relevance of this research in the field of surrogate modeling and the application of genetic algorithms is due to the growing demands on the efficiency and speed of solving complex engineering problems. This approach opens up new opportunities for optimizing the design and analysis of surrogate models, which in turn can lead to significant improvements in various industries.

2 Related Work

Scientists around the world are engaged in research and development of methods in the field of surrogate modeling, which describe effective methods for constructing simplified models (surrogates) approximating the behavior of a complex system or process.

Thus, the study [5] presents an approach to step-by-step improvement of the surrogate performance model and a new method for modeling reliability. The effectiveness of the proposed solution is confirmed by its approbation when building a reliability model for a new electrical connector.

In the article [6], a hybrid sampling method for modeling and optimizing microwave components is proposed, which selects global and local samples to form a data set using a greedy algorithm and a random sampling method. The experimental results showed that the ANN model based on the proposed approach provides higher modeling efficiency in comparison with traditional sampling methods.

The implementation of the Kriging surrogate model and the plan refinement method with a gradually expanding set of experiments on the initial model is presented in [7]. This approach is based on an evolutionary search in a neighboring sample area and has demonstrated high reliability and efficiency in real-world applications for expensive optimization.

In the framework of the study [8], an evolutionary optimizer was used to build a Kriging-based surrogate model to search for a subspace of design variables and a soft external matrix to assess the fitness function of a developing individual. In practice, this approach has shown stable optimization with a small number of experiments.

An adaptive sampling strategy based on information entropy for kriging is presented in [9]. The IE-AK algorithm made it possible to reduce the number of deterministic software calls without significant loss of accuracy while significantly saving computing resources.

In the article [10], an algorithm for global sequential sampling is proposed, which complements the experimental plan both in an unexplored area and in an area with a large error iteratively. The effectiveness of the proposed approach has been confirmed by experiments on several test functions.

Surrogate modeling of the characteristics of analog integrated circuits was carried out in the study [11]. In this article, the kriging model is studied in combination with the Latin hypercube method to evaluate the parametric performance of the circuit. The authors note that the proposed approach has proven effective in practice and can contribute to global optimization of parameters.

Thus, a review of research in the field of surrogate modeling shows that various scientists are actively developing and implementing methods to create effective and reliable surrogate models that approximate the behavior of complex systems. The presented research demonstrates a variety of approaches, including step-by-step model improvement, hybrid sampling methods, evolutionary optimizers and adaptive strategies that can significantly reduce computational resources and improve modeling accuracy.

Despite the successes achieved, there is a need for further development of surrogate modeling methods in the context of clarifying the experimental plan and applying genetic algorithms. This is due to several important aspects: an increase in the complexity of the system, the need to efficiently allocate resources by optimizing the process of selecting experimental points and improving the quality of the model, as well as the importance of implementing a flexible and adaptive approach taking into account the specifics of various tasks.

Thus, the development of a new genetic algorithm to refine the plan of surrogate modeling experiments is an urgent task that contributes to improving the quality and efficiency of modeling complex systems and processes.

3 Statement of the Problem of Surrogate Modeling

Let us consider the formulation of the surrogate modeling problem. Let $D_X \subset R^n$ – n—dimensional planning space of factors X, $\{x^{(i)}\}$, $x^{(i)} \in D_X$, $i = 1, \ldots, p$—a training data sample, $f(x^{(i)})$, $i = 1, \ldots, p$—software calculations on a high-precision model.

Then, constructing a surrogate model $\hat{f}(x)$ by approximating data pairs $\{x^{(i)}, f(x^{(i)})\}$ consists of solving the following optimization problem:

$$\sum_{i=1}^{p} L\left(\hat{f}\left(x^{(i)}, \theta\right), f\left(x^{(i)}\right)\right) \to min \quad (1)$$

where $L(.,.)$—the loss function, θ—the vector of parameters of the surrogate model.

4 Surrogate Modeling Based on Data

At the moment, the problem of reengineering software in operation is most acute due to a number of problems, including the creation of local low-resource models, increasing the speed of calculations, as well as solving import substitution problems. To build a surrogate model approximating the source software, the following approaches can be used:

1. ***Projection surrogate models (Model driven)***: Uses a top-down approach, applying available mathematical models to obtain approximations. In this case, information about the application is needed and, therefore, the approach is individual for the task.
2. ***Surrogate models based on data (Data driven)***: Assumes that absolutely nothing is known about the internal work of the software. In this case, there is a "black box" inside the software and information is collected based on empirical

approximations of the calculation results based on the original model. Since the approach is very general, it is not tied to a specific area.

3. ***Hybrid Models***: An approach to incorporate domain knowledge into a data-driven process to improve traceability and achieve accuracy with fewer parameters and estimates.

Within the framework of this study, we will rely on the fact that the initial model of the industry software is unknown or distorted and we will consider methods for constructing surrogate models based on data.

Building surrogate models on data has several key features:

1. *Data dependence*: Surrogate models are built on collected data from real systems and installations, which makes them sensitive to the quality and quantity of this data. Insufficient data or the presence of noise can negatively affect the accuracy of the model.
2. *Handling unstructured data*: In some cases, the data may be unstructured or have high dimensionality, which requires the use of preprocessing and dimensionality reduction techniques.
3. *Selection of training methods*: There are many machine learning algorithms such as regression, decision trees, neural networks, etc., and the choice of a suitable method depends on the nature of the data and the problem.
4. *Overfitting and generalizability*: It is necessary to monitor the overfitting of the model so that it can generalize and apply the learned knowledge to new data, rather than simply memorizing the training set.
5. *Interpretability*: It is important for the model to be interpretable, especially in critical domains such as medicine or finance, where understanding the decision-making process is of great importance.
6. *Adaptability*: Surrogate models must be able to adapt to changes in the data, which implies the need for regular updating and retraining of the models.
7. *Validation and testing*: It is imperative to validate the model on a holdout sample to assess its performance and accuracy.
8. *Accounting for uncertainty*: It is important to account for the uncertainties associated with the data and the model to adequately assess its robustness and stability.

The highlighted features determine the approaches and strategies used to create effective surrogate models on the data.

The process of constructing a data-driven surrogate typically consists of the following main steps:

1. *Design of Experiments (DoE)*, i.e., placing training samples in a design space relevant to the problem at hand. By design space we mean the feature space for the surrogate model (input parameters for the industry-specific software) and their domain boundaries. The strategy for sample placement depends on the source of the training data (various space-filling schemes are commonly used [12]) and the available computational budget.

2. *Collection of training data.* In most cases, training data is obtained from simulation models, as well as by directly collecting the output information from the original industry-specific software model by substituting the results of placing training samples in the design space.
3. *Model Identification.* The model parameters are found by solving an appropriately formulated minimization problem, e.g. for kriging or neural networks, or analytically, by solving an appropriate linear regression problem.
4. *Model Validation.* At this stage, the predictive ability of the model is tested, including its approximation ability (correspondence to the training data set) and generalization ability (correspondence to projects not noticed at the identification stage). In practice, it is necessary to find a balance between model approximation and generalization [13, 14].

A more detailed scheme for conducting surrogate modeling on data is presented in Fig. 1.

This study is aimed at improving existing methods for forming a plan of computational experiments in order to optimize the process of constructing a surrogate model and improve its quality.

The use of genetic algorithms in combination with machine learning methods will meet the requirements for efficiency and speed of solving complex engineering problems. The main idea of using an evolutionary approach to find optimal points in the parameter space of machine learning models is to estimate the largest surrogate error in accordance with the plan of a computational experiment.

5 Algorithms for Constructing a Plan for a Computational Experiment

Experiment planning is based on data sampling algorithms for creating a pseudo-random sample of the original data in order to improve the accuracy of the surrogate model. Let's consider the basic methods of planning a computational experiment:

1. ***Random Sampling***

   The random sampling method consists in randomly selecting points in the factor space. This approach allows us to study the behavior of a model or system without making preliminary assumptions about its structure.

   The advantage of the algorithm is that the evaluation is performed as if a simple random sample had been collected. However, with this approach to DoE, a dense distribution of points around one area and empty sectors in another area is possible, which can significantly reduce the quality of the surrogate model.
2. ***Full-factorial Sampling***

   The full-factorial sampling algorithm involves exploring all possible combinations of factor levels, which allows one to obtain a complete picture of the interactions between factors.

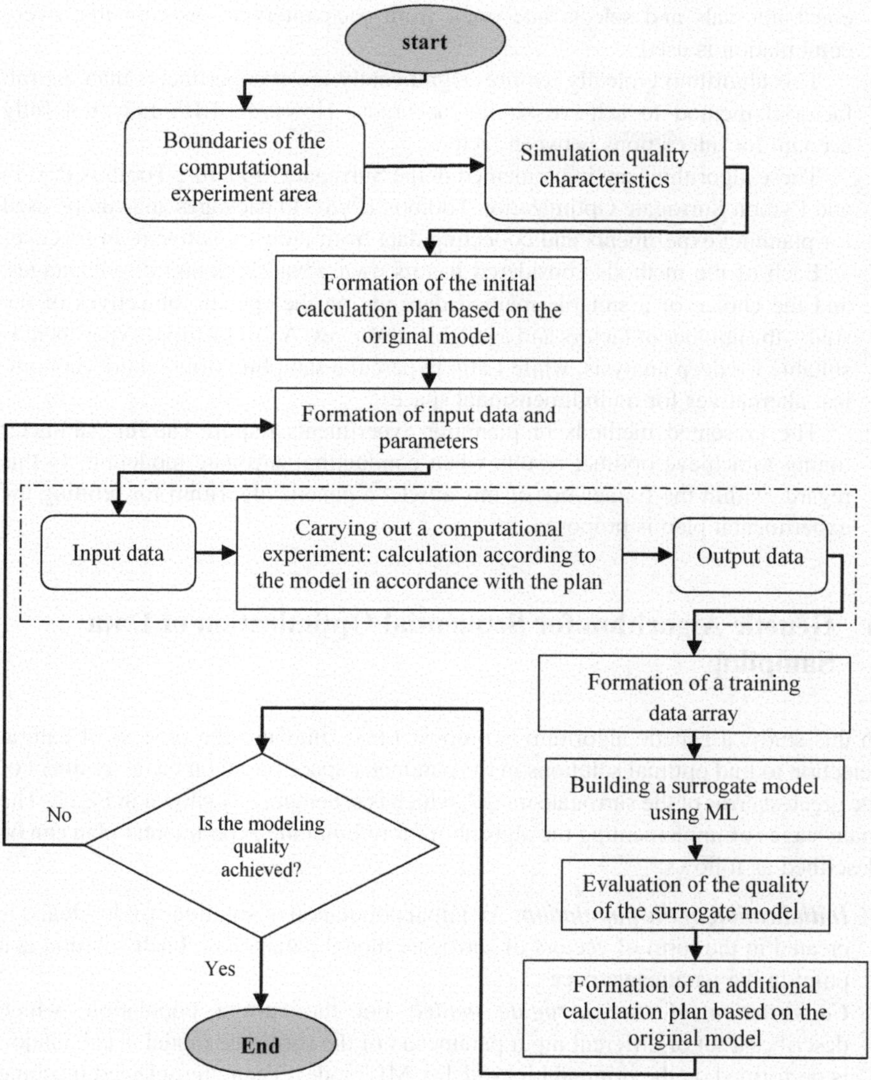

Fig. 1 Scheme for conducting surrogate modeling on data

The advantage of the algorithm is a more uniform grid in the design space, relative to simple random sampling. However, to achieve a given accuracy of the surrogate model, some areas need to be specified more densely, and for this approach this means increasing the density of the entire grid and increasing computational costs.

3. ***Latin Hypercube Sampling***

The Latin hypercube sampling (LHS) algorithm ensures that samples are distributed evenly across each factor. It divides the range of each factor into equal intervals and selects one point from each interval, ensuring that every combination is used.

This algorithm typically requires significantly fewer experiments than the full factorial method to achieve similar accuracy. However, LHS may not fully account for interactions between factors.

These algorithms are implemented in the Surrogate Modeling Toolbox (SMT) and Python Surrogate Optimization Toolbox (PySOT) packages and can be used for planning experiments and collecting data from industry software in practice.

Each of the methods considered has its own advantages and disadvantages, and the choice of a suitable method depends on the specific objectives of the study, the number of factors and available resources. A full factorial experiment is suitable for deep analysis, while Latin hypercube sampling offers more economical alternatives for multidimensional space.

The presented methods of planning experiments require careful parameter tuning to achieve optimal results when conducting surrogate modeling. In this regard, within the framework of this article, a genetic algorithm for refining the experimental plan is proposed.

6 Genetic Algorithm for Sequential Optimization of Data Sampling

In this study, a genetic algorithm is proposed that simulates the process of natural selection to find optimal solutions in the parameter space based on the assessment of the greatest error of the surrogate model, which is schematically shown in Fig. 2. The main stages of implementing the algorithm for refining the experimental plan can be described as follows:

1. ***Initialization of the population***: An initial population of solutions (individuals) is created in the form of vectors of surrogate model parameters. Each solution is a point in the parameter space.
2. ***Construction of the surrogate model***: For the current population, which describes a set of different input parameters of the surrogate model, a calculation is performed on the original high-fidelity ML model. Then, the selected machine learning model is trained to implement the stated goal of surrogate modeling.
3. ***Fitness assessment***: For each individual in the population, we assess the "fitness" depending on the largest error of the surrogate model. For example, you can use the error value as a metric for assessment, where a smaller error corresponds to greater fitness.
4. ***Selection***: It is necessary to select the best individuals for reproduction. Selection methods can include:

    - Tournament selection: several random individuals are selected, and the best one becomes a parent.
    - Roulette selection: the probability of selecting individuals is proportional to their fitness.

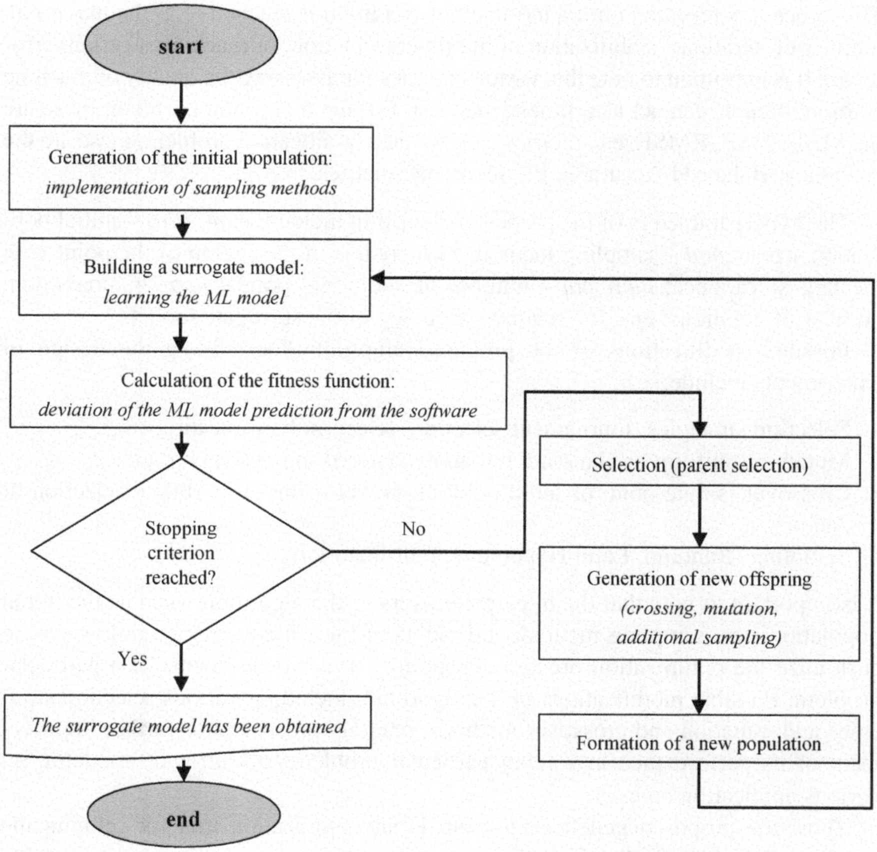

Fig. 2 Genetic algorithm for sequential optimization of data sampling for the construction of surrogate models

5. ***Crossover***: Randomly combines the parameters of two or more parents to create new offspring:
   - Single-point crossover: a point is selected and the parameters of the parents are exchanged at this point.
   - Multi-point crossover: more complex strategies for combining parameters.

6. ***Mutation***: Using a small probability, one or more parameters of the offspring are changed. Mutation helps to avoid local extremes, adding diversity to the population.
7. ***New population creation***: New offspring are generated using available sampling methods in a given area of the point with the largest deviation. The selected parents and the resulting offspring are then combined to form a new population. This may involve replacing the entire population or only a part of it, keeping the best individuals.

This process is repeated until a termination condition is reached (e.g. the maximum number of iterations, stabilization of the fitness function, or reaching a certain error level). It is important to note that various metrics for assessing the quality of machine learning models can act as a fitness function. For the regression problem, these are the MAE, MSE, RMSE, etc. metrics, and for the classification problem, these are the Accuracy, Balanced Accuracy, F1-Score, etc. metrics.

The hyperparameters of the proposed algorithm include: *num_start*—initial population size; *sempl*—sampling method; *rad*—radius of the region of the point with the largest variance; *num_pop*—number of additional points; *num_it*—maximum number of iterations; *eps_it*—required accuracy of the surrogate model.

Possible modifications of the proposed algorithm for refining the design of experiments include:

1. Selection strategies (tournament selection, selection by truncation);
2. Mutation (uniform or Gaussian mutation centered in the search region);
3. Crossover (single-point or multi-point crossover using proportional selection or "parts");
4. Sampling (Random, Latin Hypercube, Full-factorial).

It is important to note that the hyperparameters of the algorithm, such as the initial population size, sampling method, and radius of the variance region, allow you to customize the optimization process depending on the requirements of a particular problem. Possible modifications of the algorithm, including various selection strategies and mutation and crossover methods, open up prospects for further improvement of its performance in solving particular problems of surrogate modeling in various application areas.

Thus, the proposed genetic algorithm is an optimization tool for refining the experimental plan, allowing one to find optimal solutions in complex multidimensional spaces with high efficiency.

7 Computational Experiments

To implement numerical experiments, test optimization functions were selected to construct a surrogate—the Eckley function and the Rastrigin function. These functions have many local minima, which makes them difficult to optimize, and also have periodic components due to the cosine term, which adds complexity to the search for a global minimum. The global minimum of these functions is at the point $(0, 0)$, where $f(0, 0) = 0$.

The Eckley function has the following form:

$$f(x, y) = -20e^{-0.2\sqrt{0.5(x^2+y^2)}} - e^{0.5(cos(2\pi x)+cos(2\pi y))} + e + 20. \qquad (2)$$

A visualization of the Eckley function of two variables is shown in Fig. 3.

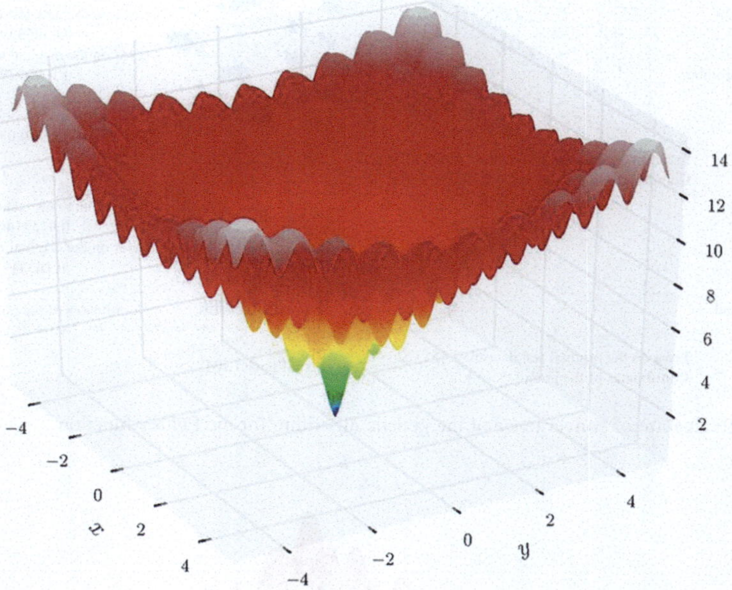

Fig. 3 Graph of the Eckley function of two variables

The results of a computational experiment on the convergence of the algorithm for the Ackley function with different configurations of the sampling method (Latin Hypercube Sampling, Full-factorial Sampling), limiting the number of iterations to 250 and stabilizing the fitness function over 10 epochs are shown in Fig. 4.

The RandomForestRegressor machine learning model is used as a surrogate model. This model is based on an ensemble approach using multiple decision trees to improve prediction accuracy and reduce the risk of overfitting. It is resistant to data noise, can handle gaps and high dimensionality, requires minimal preprocessing, and allows interpretation of the importance of features. The mean squared error (MSE) regression loss metric was used as a fitness function of the genetic algorithm.

Thus, the genetic algorithm for automating the process of generating an experimental plan demonstrated high efficiency on various configurations of sampling methods and comparable results in terms of accuracy. Despite the different experimental design (the plans for calculating the software and constructing the training set differ significantly), the mean square error on the initial test set is no more than 0.1.

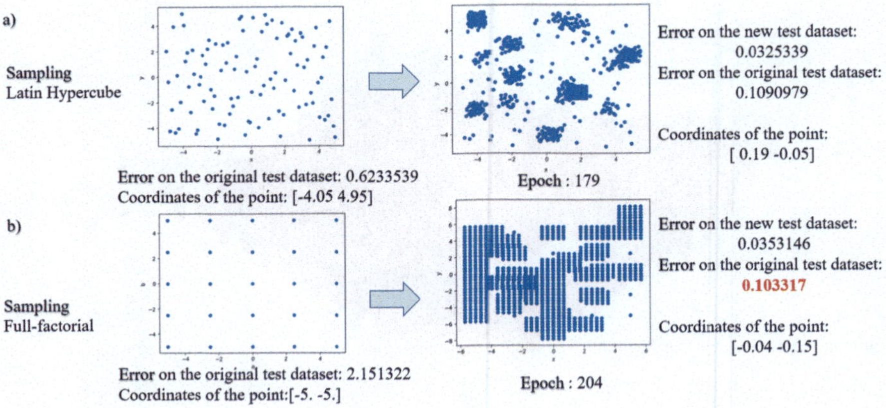

Fig. 4 Evaluation of convergence of the genetic algorithm for the Eckley function

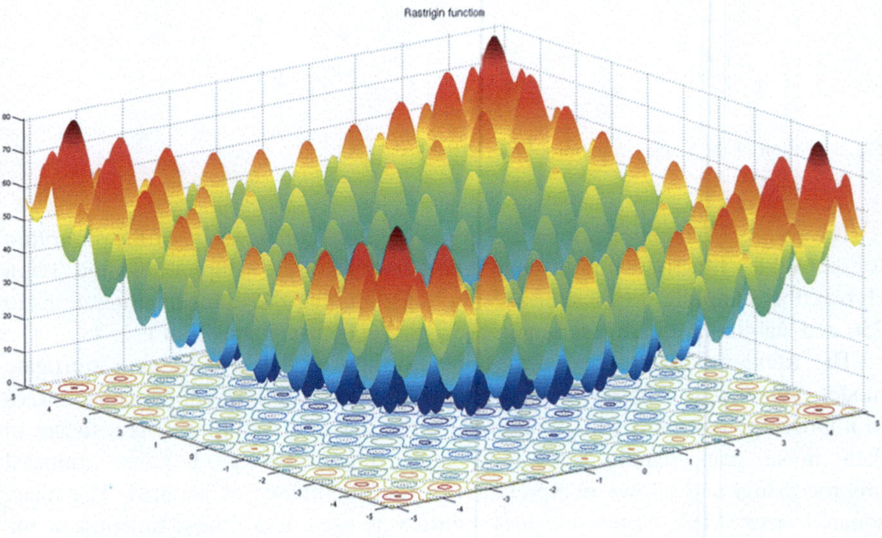

Fig. 5 Graph of the Rastrigin function of two variables

The Rastrigin function has the following form:

$$f(x_1, x_2) = An + \sum_{i=1}^{n} \left[x_i^2 - A\ \cos(2\pi x_i)\ \right],\quad \text{where } A = 10, n = 2. \tag{3}$$

A visualization of the Rastrigin function of two variables is shown in Fig. 5.

A computational experiment on the convergence of the algorithm for the Rastrigin function with different configurations of the sampling method (Latin Hypercube Sampling, Full-factorial) and a limitation of the number of iterations to

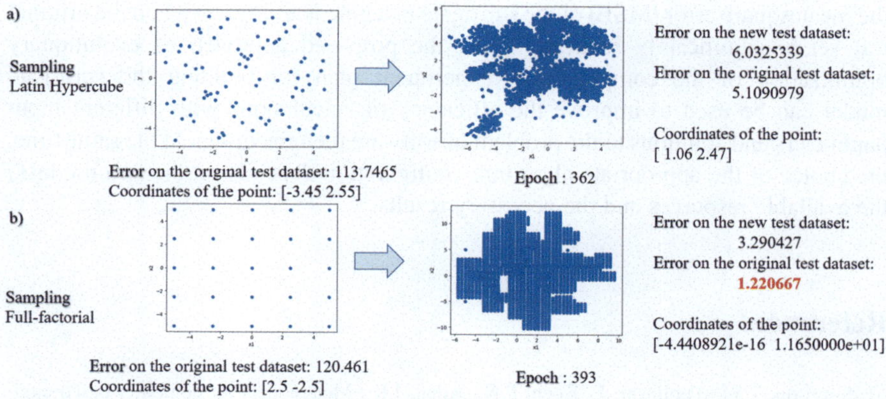

Fig. 6 Convergence evaluation of the genetic algorithm for the Rastrigin function. (**a**) Sampling Latin hypercube. (**b**) Sampling full-factorial

500 until the best fitness function stabilizes over 10 epochs confirmed the previously obtained conclusions. Note that the coordinates of the extremum points differ significantly for different sampling methods: based on the error values on the original test data set, the first configuration of the algorithm (see Fig. 6a) fell further into the local extremum than the second (see Fig. 6b). The minimum mean square error MSE on the original test set was approximately 1.2.

Thus, the proposed approach of evolutionary optimization of the computational experiment plan for building a surrogate model can be used to increase the efficiency of calculations with various input parameters and solve the problem of software reengineering. The evolutionary optimization strategy based on an adaptive approach to planning computational experiments makes it possible to find optimal model parameters and significantly reduce calculation time, which is critically important in conditions of limited resources. The prospects for further research imply a deeper study of various algorithm configurations and their adaptation to specific tasks.

8 Conclusion

The paper considers the issues of increasing the effectiveness of surrogate modeling based on sequential optimization of data sampling. The use of genetic algorithms in combination with machine learning methods is proposed to meet the requirements for efficiency and speed of solving complex engineering problems. The developed genetic algorithm for automating the process of generating an experiment plan has demonstrated high efficiency on various configurations of sampling methods and results comparable in accuracy. Despite the different design of the experiment (the plans for calculating the software and building the training set differ significantly), the mean square error (MSE) of the surrogate machine learning model on the original test set is significantly reduced. Thus, the proposed approach of evolutionary optimization of the computational experiment plan for building the surrogate model can be used to improve the efficiency of calculations with different input parameters and solutions to the problem of software reengineering. At the same time, the choice of the appropriate algorithm configuration depends on the specific task, the available resources and the necessary results.

References

1. Simpson, T.W., Peplinski, J., Koch, P.N., Allen, J.K.: Metamodels for computer-based engineering design: survey and recommendations. Comp. Simulat. **17**(2), 129–150 (2001)
2. Luo, X., Liu, P., Xia, Q., Cheng, Q., Liu, W., Mai, Y., Zhou, C., Zheng, Y., Wang, D.: Machine learning-based surrogate model assisting stochastic model predictive control of urban drainage systems. J. Environ. Manag. **346**, 118974 (2023)
3. Ren, P., Stuart, C., Zhang, M., Inomata, R., Nakamura, K., Morita, I., Spence, S.: Investigation of the surrogate model in an ANN-Meanline hybrid model for radial turbine performance prediction. Int. J. Gas Turbine Propuls. Power Syst. **15**(2), 9–18 (2024)
4. Trushin, S.M., Titlyanov, A.K.: Application of evolutionary modeling and genetic algorithms for solving optimization problems. Scient. Aspect. **46**(4), 6138–6149 (2024)
5. Liu, H., Zhang, Q., Li, X., Yu, L., Yuan, J., Kang, R.: Reliability modelling method based on the stepwise refined performance surrogate model. In: 2023 Global Reliability and Prognostics and Health Management Conference, pp. 1–7, Hangzhou, China (2023)
6. Zhang, Z., Cheng, Q.S., Chen, H., Jiang, F.: An efficient hybrid sampling method for neural network-based microwave component modeling and optimization. IEEE Microwave Wireless Compon. Lett. **30**(7), 625–628 (2020)
7. Suprayitno, Yu J.-C.: Evolutionary algorithm using progressive kriging model and dynamic reliable region for expensive optimization problems. In: IEEE International Conference on Systems, Man, and Cybernetics, pp. 004383–004388. Budapest, Hungary (2016)
8. Yu, J.-C.: Suprayitno: evolutionary reliable regional kriging surrogate and soft outer array for robust engineering optimization. IEEE Access. **5**, 16520–16531 (2017)
9. Jin, Z., Li, J.: IE-AK: a novel adaptive sampling strategy based on information entropy for kriging in metamodel-based reliability analysis. Reliab. Eng. Syst. Saf. **229**(4), 108824 (2022)
10. Wang, X., Song, B., Wang, P.: Surrogate-based global sequential sampling algorithm. In: 9th International Symposium on Computational Intelligence and Design, pp. 122–125, Hangzhou, China (2016)
11. You, H., Yang, M., Wang, D., Jia, X.: Kriging model combined with latin hypercube sampling for surrogate modeling of analog integrated circuit performance. In: 10th International Symposium on Quality Electronic Design, pp. 554–558, San Jose, CA (2009)
12. Simpson, T.W., Peplinski, J., Koch, P.N., Allen, J.K.: Metamodels for computer-based engineering design: survey and recommendations. Eng. Comput. **17**, 129–150 (2001)
13. Arlot, S., Celisse, A.: A survey of cross-validation procedures for model selection. Statist. Surv. **4**, 40–79 (2010)
14. James, G., Witten, D., Hastie, T., Tibshirani, R.: An Introduction to Statistical Learning. Springer, New York, NY (2013)

Beltrami-Mitchell Equations for the Case of Incompatible Deformations

K. N. Pestov, M. A. Guzev, and O. N. Lyubimova

Abstract In this paper, a generalization and a covariant form of the Beltrami-Mitchell equations are derived for the case of incompatible deformations. The representation of the internal material metric and the associated coefficients of the non-Euclidean connectivity are given. A relation between them and the Euclidean connectivity coefficients defined by the deformation properties of the material is obtained. An incompatibility condition in covariant form is presented. The paper introduces an additional intrinsic non-Euclidean material characterization in terms of the Ricci tensor. The resulting Beltrami-Mitchell equations are given in Cartesian rectangular and cylindrical coordinate systems. When considering the scheme of the non-Euclidean model, the Ricci tensor is emphasized as a geometric parameter responsible for inelastic deformation. The analytical form of an additional internal non-Euclidean material characteristic for the problem of the plane deformed state in an axisymmetric cylinder is given. The analytical solution of the Beltrami-Mitchell equations for the problem of the plane deformed state in an axisymmetric cylinder is obtained.

Keywords Beltrami-Mitchell equations · Saint-Venant consistency conditions · Non-Euclidean continuum model · Ricci tensor

K. N. Pestov (✉)
Vladivostok Branch of Russian Customs Academy, Vladivostok, Russia

M. A. Guzev
Institute for Applied Mathematics Far Eastern Branch of Russian Academy Sciences, Vladivostok, Russia

O. N. Lyubimova
Far Eastern Federal University, Vladivostok, Russia

A. Gibadullin, S. Gordin (eds.), *Computing Technologies and Applied Mathematics*, Springer Proceedings in Mathematics & Statistics 500,
https://doi.org/10.1007/978-3-031-90546-9_9

1 Introduction

In the theory of elasticity, the classical tensor objects that describe the state of a continuous medium include the displacement vector, deformation tensor, and stress tensor. However, when considering various types of defects, additional tensor objects may need to be introduced.

In the field of continuum mechanics, the interest in the geometric and physical interpretation of tensor objects that determine the local properties of a variety remains relevant [1, 2]. Such objects include the Riemann curvature tensor, nonmetricity, and the torsion tensor of connectivity.

In [3], based on the calibration approach, which has proven its effectiveness in modeling in elementary particle physics, the dynamics of defects is constructed. The geometric interpretation of the calibration approach is that the calibration fields are connections in the main bundle above the base. The base is an area occupied by a continuous medium with a structural group that is a semi-direct product of a group of translations and a group of rotations of three-dimensional space.

In [4], defects such as dislocations and disclinations are described within the framework of Riemann-Cartan geometry using the torsion tensor and the curvature tensor.

In [5], variants of non-Euclidean continuum models are proposed, considering the interaction of various defective structures.

Within the framework of linear approximation of deformations, the non-trivial components of the Riemann tensor are equivalent to the Saint-Venant compatibility equations [6, 7].

In three-dimensional space, the Riemann curvature tensor is completely determined by the Ricci tensor $\widehat{R}$ [8]

$$R_{ijkl} = R_{ik} g_{jl} - R_{il} g_{jk} + R_{jl} g_{ik} - R_{jk} g_{il} + \frac{R}{2}\left(g_{il} g_{jk} - g_{ik} g_{jl}\right) \qquad (1)$$

where $R = Tr\left(\widehat{R}\right) = R_{mn} g^{mn}$—trace of the Ricci tensor (scalar curvature), g_{jk}—components of the metric tensor $\widehat{g}$, $i,j,k,l = 1,2,3$.

In three-dimensional space, the independent components of the Riemann curvature tensor are only 6 out of 81. The Ricci tensor obtained by convolution of two indices in the curvature tensor is a symmetric tensor of the second rank, which also has 6 independent components in three-dimensional space. Even though both tensors are zero for classical models, the expression of the Ricci tensor in deformations or stresses can be useful when extending these models to Riemannian manifolds.

The Beltrami-Mitchell equations, written as [9]

$$\Delta\widehat{\sigma} + \frac{1}{1+\nu} \nabla\nabla Tr(\widehat{\sigma}) = -\nabla\vec{F} - \left(\nabla\vec{F}\right)^T - \frac{\nu}{1-\nu}\widehat{g}\nabla\cdot\vec{F} \qquad (2)$$

where ν—Poisson's ratio, $\widehat{\sigma}$—stress tensor, $\vec{F}$—vector of external volumetric forces, $\left(\nabla\vec{F}\right)^T$—the transposed tensor to $\nabla\vec{F}$ is tensor object.

With such a record of the equations, the equations can be represented as equal to zero of some symmetric tensor of the second rank. In Euclidean models of continuum mechanics, the Ricci tensor also has these properties and is a symmetric tensor of the second rank, equal to zero.

It is necessary to establish a connection between the specified tensor objects.

To do this, it is necessary to obtain an expression of the Ricci tensor in terms of deformations and stresses.

In classical textbooks such as [6, 7], the derivation of the Beltrami-Mitchell equations is carried out by differentiating, grouping and combining the equations of equilibrium, compatibility and Hooke's law. This method appears to be somewhat artificial mathematical techniques.

In [10], the Beltrami-Mitchell equations are derived for Riemann spaces by transformations of the components of the Riemann tensor. In [11], these equations are derived from the Castigliano variational principle, while their geometric interpretation is lost.

In this paper, these equations are derived formally when calculating the components of the Ricci tensor for a continuous medium in a deformed state.

It is known that in the general case, elastic deformation is incompatible. In this regard, it becomes necessary to consider how the classical Beltrami-Mitchell equations change when considering incompatible elastic deformations.

2 Connection of the Ricci Tensor with the Strain Tensor

The connection between the Ricci tensor and the strain tensor field is a crucial aspect that requires thorough examination. To fully comprehend the process of deriving the Beltrami–Mitchell equations, it is essential to systematically describe all steps involved.

Consider a continuous medium in its undeformed, or reporting, configuration. This configuration is described using a curved coordinate system, with the corresponding components of the metric tensor g_{ij} and Levi–Civita connection components Γ^i_{kl}. In this state, the Ricci tensor is assumed to be zero

$$\widehat{R}=0. \tag{3}$$

Upon deformation, the medium transitions to its deformed, or actual, configuration. The components of the metric tensor $\widetilde{g}_{ij}$ and the corresponding connection components associated with this new metric determine the components of the Ricci tensor, as expressed

$$\widetilde{R}_{jk} = \frac{\partial \widetilde{\Gamma}^i_{ji}}{\partial x^k} - \frac{\partial \widetilde{\Gamma}^i_{jk}}{\partial x^i} + \widetilde{\Gamma}^i_{km}\widetilde{\Gamma}^m_{ji} - \widetilde{\Gamma}^i_{im}\widetilde{\Gamma}^m_{jk}, \tag{4}$$

where the Levi–Civita connection components remain unchanged, as represented by equation

$$\widetilde{\Gamma}^i_{kl} = \frac{1}{2}\widetilde{g}^{im}\left(\frac{\partial \widetilde{g}_{mk}}{\partial x^l} + \frac{\partial \widetilde{g}_{ml}}{\partial x^k} - \frac{\partial \widetilde{g}_{kl}}{\partial x^m}\right). \tag{5}$$

The components of the metric tensors for the actual and reported configurations are related through the components of the strain tensor ε_{ij}

$$\widetilde{g}_{ij} = g_{ij} + 2\varepsilon_{ij}. \tag{6}$$

In linear approximation, the components of the inverse metric tensor can be expressed in terms of the actual configuration

$$\widetilde{g}^{ij} = g^{ij} - 2\varepsilon^{ij}. \tag{7}$$

By substituting expressions (6) и (7) into the (5) and considering only the first-order deformations, specific equality obtain

$$\widetilde{\Gamma}^i_{kl} = \frac{1}{2}g^{im}\left(\frac{\partial g_{mk}}{\partial x^l} + \frac{\partial g_{ml}}{\partial x^k} - \frac{\partial g_{kl}}{\partial x^m}\right) + g^{im}\left(\frac{\partial \varepsilon_{mk}}{\partial x^l} + \frac{\partial \varepsilon_{ml}}{\partial x^k} - \frac{\partial \varepsilon_{kl}}{\partial x^m}\right) -$$
$$\varepsilon^{im}\left(\frac{\partial g_{mk}}{\partial x^l} + \frac{\partial g_{ml}}{\partial x^k} - \frac{\partial g_{kl}}{\partial x^m}\right) = \Gamma^i_{kl} + g^{im}\left(\frac{\partial \varepsilon_{mk}}{\partial x^l} + \frac{\partial \varepsilon_{ml}}{\partial x^k} - \frac{\partial \varepsilon_{kl}}{\partial x^m}\right) - 2\varepsilon^{im}\Gamma_{mkl} \tag{8}$$

Using the operations of raising and lowering index

$$\varepsilon^{im}\Gamma_{mkl} = \varepsilon_{mp}g^{im}\Gamma^p_{kl}, \tag{9}$$

the expression (8) is transformed to the form

$$\widetilde{\Gamma}^i_{kl} = \Gamma^i_{kl} + g^{im}\left(\frac{\partial \varepsilon_{mk}}{\partial x^l} - \Gamma^p_{kl}\varepsilon_{mp} - \Gamma^p_{ml}\varepsilon_{pk}\right) + g^{im}\left(\frac{\partial \varepsilon_{ml}}{\partial x^k} - \Gamma^p_{kl}\varepsilon_{mp} - \Gamma^p_{km}\varepsilon_{pl}\right) -$$
$$- g^{im}\left(\frac{\partial \varepsilon_{kl}}{\partial x^m} - \Gamma^p_{km}\varepsilon_{pl} - \Gamma^p_{ml}\varepsilon_{pk}\right) \tag{10}$$

Components in brackets can be expressed in terms of the covariant derivative $\nabla_l \varepsilon_{mk}$, $\nabla_k \varepsilon_{ml}$ and $\nabla_m \varepsilon_{lk}$, which is the covariant component of a second-order tensor with respect to the Euclidean metric connection

$$\begin{aligned}\nabla_l \varepsilon_{mk} &= \frac{\partial \varepsilon_{mk}}{\partial x^l} - \Gamma^p_{kl}\varepsilon_{mp} - \Gamma^p_{ml}\varepsilon_{pk},\\ \nabla_k \varepsilon_{ml} &= \frac{\partial \varepsilon_{ml}}{\partial x^k} - \Gamma^p_{kl}\varepsilon_{mp} - \Gamma^p_{km}\varepsilon_{pl},\\ \nabla_m \varepsilon_{lk} &= \frac{\partial \varepsilon_{kl}}{\partial x^m} - \Gamma^p_{km}\varepsilon_{pl} - \Gamma^p_{ml}\varepsilon_{pk}.\end{aligned} \tag{11}$$

Since the metric is consistent with the concept of connectivity, the operation of raising the index doesn't conflict with covariant differentiation

$$g^{im}\nabla_l \varepsilon_{mk} = \nabla_l \left(g^{im}\varepsilon_{mk}\right) = \nabla_l \varepsilon^i_k. \tag{12}$$

The following expression is obtained

$$\widetilde{\Gamma}^i_{kl} = \Gamma^i_{kl} + \nabla_l \varepsilon^i_k + \nabla_k \varepsilon^i_l - g^{im}\nabla_m \varepsilon_{kl}. \tag{13}$$

By raising the index m, we get the tensor

$$\nabla^i \varepsilon_{kl} = g^{im}\nabla_m \varepsilon_{kl}. \tag{14}$$

Let's introduce a lower-index symmetric tensor

$$E^i_{kl} = \nabla_l \varepsilon^i_k + \nabla_k \varepsilon^i_l - \nabla^i \varepsilon_{kl}, \tag{15}$$

From formulas (14)–(16) it is possible to obtain the final expression for the connectivity coefficients

$$\widetilde{\Gamma}^i_{kl} = \Gamma^i_{kl} + \mathrm{E}^i_{kl} \tag{16}$$

Relation (16) can be interpreted as connectivity transformation, and in the scientific literature [12] E^i_{kl} is called the affine strain tensor.

Substituting (16) into the formula of the Ricci tensor component (4) and performing subsequent calculations, we obtain

$$\widetilde{R}_{jk}=\frac{\partial\left(\Gamma^i_{ji}+E^i_{ji}\right)}{\partial x^k}-\frac{\partial\left(\Gamma^i_{jk}+E^i_{jk}\right)}{\partial x^i}+\left(\Gamma^i_{km}+E^i_{km}\right)\left(\Gamma^m_{ji}+E^m_{ji}\right)-\left(\Gamma^m_{jk}+E^m_{jk}\right)\left(\Gamma^i_{im}+E^i_{im}\right)=$$
$$=R_{jk}+\frac{\partial E^i_{ji}}{\partial x^k}-\frac{\partial E^i_{jk}}{\partial x^i}+\Gamma^i_{km}E^m_{ji}+\Gamma^m_{ji}E^i_{km}-\Gamma^m_{jk}E^i_{im}-\Gamma^i_{im}E^m_{jk}+E^i_{km}E^m_{ji}-E^m_{jk}E^i_{im}=$$
$$=R_{jk}+\frac{\partial E^i_{ji}}{\partial x^k}-\Gamma^m_{jk}E^i_{im}-\left(\frac{\partial E^i_{jk}}{\partial x^i}-\Gamma^i_{km}E^m_{ji}-\Gamma^m_{ji}E^i_{km}+\Gamma^i_{im}E^m_{jk}\right)+E^i_{km}E^m_{ji}-E^m_{jk}E^i_{im}$$
$$=R_{jk}+\nabla_k E^i_{ji}-\nabla_i E^i_{jk}+E^i_{km}E^m_{ji}-E^m_{jk}E^i_{im}. \tag{17}$$

or considering (3) in tensor form

$$\widetilde{\widehat{R}}=\nabla Tr\left(\widehat{E}\right)-\nabla\cdot\widehat{E}+\widehat{E}\cdot\cdot\widehat{E}-\widehat{E}\cdot Tr\left(\widehat{E}\right). \tag{18}$$

Formula (18) is similar to the expression for the Riemann tensor transformation law for torsion-free metric deformation [12], limited to the first order, we have

$$\widetilde{R}_{jk}=\nabla_k \mathrm{E}^i_{ji}-\nabla_i \mathrm{E}^i_{jk} \tag{19}$$

Returning to the deformations according to formulas (16),

$$\begin{aligned}\widetilde{R}_{jk}&=\nabla_k\left(\nabla_j\varepsilon^i_i+\nabla_i\varepsilon^i_j-\nabla^i\varepsilon_{ij}\right)-\nabla_i\left(\nabla_j\varepsilon^i_k+\nabla_k\varepsilon^i_j-\nabla^i\varepsilon_{kj}\right)=\\&=\nabla_k\nabla_j\varepsilon^i_i+\nabla_k\nabla_i\varepsilon^i_j-\nabla_k\nabla^i\varepsilon_{ij}-\nabla_i\nabla_j\varepsilon^i_k-\nabla_i\nabla_k\varepsilon^i_j+\nabla_i\nabla^i\varepsilon_{kj}=\\&=\nabla_k\nabla_j\varepsilon^i_i-\nabla_i\nabla_j\varepsilon^i_k-\nabla_i\nabla_k\varepsilon^i_j+\nabla_i\nabla^i\varepsilon_{kj}.\end{aligned} \tag{20}$$

The Ricci tensor can be rewritten as

$$\widetilde{\widehat{R}}=\Delta\widehat{\varepsilon}+\nabla\nabla Tr(\widehat{\varepsilon})-\nabla(\nabla\cdot\widehat{\varepsilon})-(\nabla(\nabla\cdot\widehat{\varepsilon}))^T \tag{21}$$

The symmetry of the right side is provided by the symmetry of the strain tensor, the Euclidean geometry of the original space in which covariant derivatives commute, as well as the symmetry of the expression $\nabla(\nabla\cdot\widehat{\varepsilon})$.

3 Connection of the Ricci Tensor with the Stress Tensor

Hooke's law through the Lame coefficients can be written in tensor form

$$\widehat{\varepsilon}=\frac{1}{2\mu}\left(\widehat{\sigma}-\frac{\lambda}{3\lambda+2\mu}Tr(\widehat{\sigma})\widehat{g}\right) \tag{22}$$

Then, the Ricci tensor (21) can be expressed in terms of the stress tensor as follows:

$$\widetilde{\widehat{R}}=\frac{1}{2\mu}\left[\Delta\widehat{\sigma}+\frac{2\lambda+2\mu}{3\lambda+2\mu}\nabla\nabla Tr(\widehat{\sigma})-\frac{\lambda\Delta Tr(\widehat{\sigma})}{3\lambda+2\mu}\widehat{g}-\nabla(\nabla\cdot\widehat{\sigma})-(\nabla(\nabla\cdot\widehat{\sigma}))^{T}\right] \tag{23}$$

4 The Beltrami-Mitchell Equations

Under the conditions of equilibrium equations

$$\nabla\cdot\widehat{\sigma}+\vec{F}=0 \tag{24}$$

where $\vec{F}$ is the vector of volumetric forces, the Ricci tensor (22) is transformed into

$$\widetilde{\widehat{R}}=\frac{1}{2\mu}\left[\Delta\widehat{\sigma}+\frac{2\lambda+2\mu}{3\lambda+2\mu}\nabla\nabla Tr(\widehat{\sigma})-\frac{\lambda\Delta Tr(\widehat{\sigma})}{3\lambda+2\mu}\widehat{g}+\nabla\vec{F}+\left(\nabla\vec{F}\right)^{T}\right] \tag{25}$$

In the classical formulation, the Riemann tensor and the Ricci tensor for the configuration under consideration are zero (the continuous medium is immersed in three-dimensional Euclidean space). Using the condition that the curvature is equal to zero, it is possible to exclude $\Delta Tr(\widehat{\sigma})$ in the components of the Ricci tensor using Eq. (25).

Calculating the scalar curvature in a linear stress approximation gives the following equation

$$R=\widetilde{R}_{jk}\widetilde{g}^{jk}=\frac{1}{\mu}\left[\frac{\lambda+2\mu}{3\lambda+2\mu}\Delta Tr(\widehat{\sigma})-\nabla\cdot(\nabla\cdot\widehat{\sigma})\right]=0 \tag{26}$$

then we get

$$\Delta Tr(\widehat{\sigma})=\frac{3\lambda+2\mu}{\lambda+2\mu}\nabla\cdot(\nabla\cdot\widehat{\sigma}). \tag{27}$$

Substituting (28) into (26), it turns out

$$\widetilde{\widehat{R}}=\frac{1}{2\mu}\left[\Delta\widehat{\sigma}+\frac{2\lambda+2\mu}{3\lambda+2\mu}\nabla\nabla Tr(\widehat{\sigma})+\nabla\vec{F}+\left(\nabla\vec{F}\right)^{T}+\frac{\lambda\nabla\cdot\vec{F}}{\lambda+2\mu}\widehat{g}\right]. \tag{28}$$

The Ricci tensor (28) is zero, which corresponds to the Beltrami-Mitchell equations

$$\Delta\widehat{\sigma}+\frac{2\lambda+2\mu}{3\lambda+2\mu}\nabla\nabla Tr(\widehat{\sigma})+\nabla\vec{F}+\left(\nabla\vec{F}\right)^{T}+\frac{\lambda\nabla\cdot\vec{F}}{\lambda+2\mu}\widehat{g}=0, \tag{29}$$

If we rewrite Eq. (29) using standard technical constants E, ν, we get Eq. (2).

Thus, the components of the Ricci tensor in the linear theory of stress elasticity are nonzero components of the Beltrami-Mitchell equations.

5 Incompatible Deformations

Incompatible deformations occur when Saint-Venant conditions for deformation components are violated

$$\nabla_n\nabla_m\varepsilon_{pq}+\nabla_p\nabla_q\varepsilon_{mn}-\nabla_m\nabla_q\varepsilon_{pn}-\nabla_p\nabla_n\varepsilon_{mq}=0, \tag{30}$$

This is equivalent to the equality of the Riemann tensor to zero. In such a situation, instead of using the Riemann tensor, which has 81 components and only 6 independent ones, it is more appropriate to use the Ricci tensor or the Einstein tensor to characterize the incompatibility of deformations

$$\widehat{G}=\widehat{R}-\frac{1}{2}R\widehat{g} \tag{31}$$

Let $\widehat{R}^{(u)}$ is the Ricci tensor, which characterizes the tensor of small incompatible deformations $\widehat{\varepsilon}^{(u)}$. Let $\widehat{R}^{(e)}$ is the Ricci tensor of elastic deformations $\widehat{\varepsilon}^{(e)}$. Then the tensor of total deformations can be represented as a sum:

$$\widehat{\varepsilon}=\widehat{\varepsilon}^{(e)}+\widehat{\varepsilon}^{(u)},$$

considering the linearity of (21) and (22), the Ricci tensor of their sum is

$$\widehat{R}=\widehat{R}^{(e)}+\widehat{R}^{(u)}=0. \tag{32}$$

Using the stress expression according to formula (28), equality (32) can be represented as follows

$$\Delta\widehat{\sigma}+\frac{2\lambda+2\mu}{3\lambda+2\mu}\nabla\nabla Tr(\widehat{\sigma})-\frac{\lambda\Delta Tr(\widehat{\sigma})}{3\lambda+2\mu}\widehat{g}+\nabla\vec{F}+\left(\nabla\vec{F}\right)^{T}+2\mu\widehat{R}^{(u)}=0. \tag{33}$$

In this case, similarly to obtaining classical equations, it is possible to exclude the $\Delta Tr(\widehat{\sigma})$ summand by taking the trace of the matrix from the expression (33).

$$\frac{\lambda+2\mu}{3\lambda+2\mu}\Delta Tr(\widehat{\sigma})+\nabla\cdot\vec{F}+\mu R^{(u)}=0 \tag{34}$$

where $R^{(u)}=Tr\left(\widehat{R}^{(u)}\right)$. As a result, (33) can be represented as follows

$$\Delta\widehat{\sigma}+\frac{2\lambda+2\mu}{3\lambda+2\mu}\nabla\nabla Tr(\widehat{\sigma})+\nabla\vec{F}+\left(\nabla\vec{F}\right)^{T}+\frac{\lambda\nabla\cdot\vec{F}}{\lambda+2\mu}\widehat{g}+2\mu\widehat{R}^{(u)}+\frac{\lambda\mu R^{(u)}}{\lambda+2\mu}=0, \tag{35}$$

Equations (36) are an analogue of the Beltrami-Mitchell equations for an elastic medium with incompatible deformations, which are characterized by the Ricci tensor$\widehat{R}^{(u)}$. If incompatible deformations are characterized by the Einstein tensor $\widehat{G}^{(e)}$, which, as noted in [13], coincides with the incompatibility tensor, the components of the Ricci tensor$\widehat{R}^{(u)}$ in (36) can be expressed in $\widehat{G}^{(e)}$, using the definition (32) and its trace:

$$Tr\left(\widehat{G}^{(u)}\right)=-\frac{1}{2}R^{(u)}, \tag{36}$$

then

$$\widehat{R}^{(u)}=\widehat{G}^{(u)}+G^{(u)}\widehat{g}, \tag{37}$$

where $G^{(u)}=Tr\left(\widehat{G}^{(u)}\right)$.

Substituting (36) and (37) into the (35), we obtain:

$$\Delta\widehat{\sigma}+\frac{2\lambda+2\mu}{3\lambda+2\mu}\nabla\nabla Tr(\widehat{\sigma})+\nabla\vec{F}+\left(\nabla\vec{F}\right)^{T}+\frac{\lambda\nabla\cdot\vec{F}}{\lambda+2\mu}\widehat{g}=2\mu\left(\widehat{G}^{(u)}+G^{(u)}\widehat{g}\right)-\frac{2\lambda\mu G^{(u)}}{\lambda+2\mu}, \tag{38}$$

Equations (38) are an analogue of the Beltrami-Mitchell equations for an elastic medium with incompatible deformations, which are characterized by the Einstein tensor $\widehat{G}^{(u)}$.

In the Cartesian coordinate system (x, y, z) Lame parameters $H_x=H_y=H_z=1$, the connectivity coefficients are zero and Eq. (35) take the simplest form

$$
\begin{aligned}
&\Delta\sigma_{xx}+\frac{2\lambda+2\mu}{3\lambda+2\mu}\frac{\partial^2\sigma}{\partial x^2}+2\frac{\partial F_x}{\partial x}+\frac{\lambda}{\lambda+2\mu}\left(\frac{\partial F_x}{\partial x}+\frac{\partial F_y}{\partial y}+\frac{\partial F_z}{\partial z}\right)=2\mu R^{(u)}_{xx}+\frac{\lambda\mu R^{(u)}}{\lambda+2\mu},\\
&\Delta\sigma_{yy}+\frac{2\lambda+2\mu}{3\lambda+2\mu}\frac{\partial^2\sigma}{\partial y^2}+2\frac{\partial F_y}{\partial y}+\frac{\lambda}{\lambda+2\mu}\left(\frac{\partial F_x}{\partial x}+\frac{\partial F_y}{\partial y}+\frac{\partial F_z}{\partial z}\right)=2\mu R^{(u)}_{yy}+\frac{\lambda\mu R^{(u)}}{\lambda+2\mu},\\
&\Delta\sigma_{zz}+\frac{2\lambda+2\mu}{3\lambda+2\mu}\frac{\partial^2\sigma}{\partial z^2}+2\frac{\partial F_z}{\partial z}+\frac{\lambda}{\lambda+2\mu}\left(\frac{\partial F_x}{\partial x}+\frac{\partial F_y}{\partial y}+\frac{\partial F_z}{\partial z}\right)=2\mu R^{(u)}_{zz}+\frac{\lambda\mu R^{(u)}}{\lambda+2\mu},\\
&\Delta\sigma_{xy}+\frac{2\lambda+2\mu}{3\lambda+2\mu}\frac{\partial^2\sigma}{\partial x\partial y}+\frac{\partial F_x}{\partial y}+\frac{\partial F_y}{\partial x}=2\mu R^{(u)}_{xy},\\
&\Delta\sigma_{xz}+\frac{2\lambda+2\mu}{3\lambda+2\mu}\frac{\partial^2\sigma}{\partial x\partial z}+\frac{\partial F_x}{\partial z}+\frac{\partial F_z}{\partial x}=2\mu R^{(u)}_{xz},\\
&\Delta\sigma_{yz}+\frac{2\lambda+2\mu}{3\lambda+2\mu}\frac{\partial^2\sigma}{\partial y\partial z}+\frac{\partial F_z}{\partial y}+\frac{\partial F_y}{\partial z}=2\mu R^{(u)}_{yz},
\end{aligned}
\tag{39}
$$

here $\sigma = \sigma_{xx} + \sigma_{yy} + \sigma_{zz}$, $R^{(u)} = R^{(u)}_{xx} + R^{(u)}_{yy} + R^{(u)}_{zz}$ and the Laplace operator $\Delta = \frac{\partial^2}{\partial x^2} + \frac{\partial^2}{\partial y^2} + \frac{\partial^2}{\partial z^2}$.

In a cylindrical coordinate system (r, ϕ, z) the parameters of the Lame are equal $H_r = H_z = 1$, $H_\phi = r$, non-zero connectivity components $\Gamma^r_{\phi\phi} = -r$, $\Gamma^\phi_{r\phi} = \frac{1}{r}$ and the Eq. (35) will take the form

$$
\begin{aligned}
&\Delta\sigma_{rr}-\frac{2}{r^2}(\sigma_{rr}-\sigma_{\phi\phi})-\frac{4}{r^2}\frac{\partial\sigma_{r\phi}}{\partial\phi}+\frac{2\lambda+2\mu}{3\lambda+2\mu}\frac{\partial^2\sigma}{\partial r^2}+2\frac{\partial F_r}{\partial r}++\frac{\lambda}{\lambda+2\mu}\\
&\times\left(\frac{\partial F_r}{\partial r}+\frac{1}{r}\frac{\partial F_\phi}{\partial\phi}+\frac{F_r}{r}+\frac{\partial F_z}{\partial z}\right)+2\mu R^{(u)}_{rr}+\frac{\lambda\mu R^{(u)}}{\lambda+2\mu}=0,\ \Delta\sigma_{\phi\phi}+\frac{2}{r^2}(\sigma_{rr}-\sigma_{\phi\phi})\\
&+\frac{4}{r^2}\frac{\partial\sigma_{r\phi}}{\partial\phi}+\frac{2\lambda+2\mu}{3\lambda+2\mu}\left(\frac{1}{r}\frac{\partial\sigma}{\partial r}+\frac{1}{r}\frac{\partial^2\sigma}{\partial\phi^2}\right)++2\left(\frac{1}{r}\frac{\partial F_\phi}{\partial\phi}+\frac{F_r}{r}\right)+\frac{\lambda}{\lambda+2\mu}\\
&\times\left(\frac{\partial F_r}{\partial r}+\frac{1}{r}\frac{\partial F_\phi}{\partial\phi}+\frac{F_r}{r}+\frac{\partial F_z}{\partial z}\right)+2\mu R^{(u)}_{\phi\phi}+\frac{\lambda\mu R^{(u)}}{\lambda+2\mu}=0,\ \Delta\sigma_{zz}+\frac{2\lambda+2\mu}{3\lambda+2\mu}\frac{\partial^2\sigma}{\partial z^2}\\
&+2\frac{\partial F_z}{\partial z}+\frac{\lambda}{\lambda+2\mu}\left(\frac{\partial F_r}{\partial r}+\frac{1}{r}\frac{\partial F_\phi}{\partial\phi}+\frac{F_r}{r}+\frac{\partial F_z}{\partial z}\right)+2\mu R^{(u)}_{zz}++\frac{\lambda\mu R^{(u)}}{\lambda+2\mu}=0,
\end{aligned}
$$

$$
\begin{aligned}
&\Delta\sigma_{r\phi}+\frac{2\lambda+2\mu}{3\lambda+2\mu}\frac{\partial}{\partial r}\left(\frac{1}{r}\frac{\partial\sigma}{\partial\phi}\right)+\frac{2}{r^2}\frac{\partial}{\partial\phi}(\sigma_{rr}-\sigma_{\phi\phi})-\frac{\sigma_{\phi z}}{r^2}+\frac{\partial F_\phi}{\partial r}+\frac{1}{r}\frac{\partial F_r}{\partial\phi}-\frac{F_\phi}{r}\\
&+2\mu R^{(u)}_{r\phi}=0,
\end{aligned}
$$

$$
\Delta\sigma_{\phi z}+\frac{2\lambda+2\mu}{3\lambda+2\mu}\frac{1}{r}\frac{\partial^2\sigma}{\partial\phi\partial z}+\frac{2}{r^2}\frac{\partial\sigma_{rz}}{\partial\phi}-\frac{\sigma_{\phi z}}{r^2}+\frac{1}{r}\frac{\partial F_z}{\partial\phi}+\frac{\partial F_\phi}{\partial z}+2\mu R^{(u)}_{xz}=0,
$$

$$\Delta\sigma_{rz} + \frac{2\lambda+2\mu}{3\lambda+2\mu}\frac{\partial^2\sigma}{\partial r\partial z} - \frac{2}{r^2}\frac{\partial\sigma_{rz}}{\partial\phi} - \frac{\sigma_{rz}}{r^2} + \frac{\partial F_z}{\partial r} + \frac{\partial F_r}{\partial z} + 2\mu R^{(u)}_{yz} = 0,$$

where $\sigma = \sigma_{rr} + \sigma_{\phi\phi} + \sigma_{zz}$, $R^{(u)} = R^{(u)}_{rr} + R^{(u)}_{\phi\phi} + R^{(u)}_{zz}$ and the Laplace operator $\Delta = \frac{\partial^2}{\partial r^2} + \frac{1}{r}\frac{\partial}{\partial r} + \frac{1}{r^2}\frac{\partial^2}{\partial\phi^2} + \frac{\partial^2}{\partial z^2}$.

6 Example

Let us consider the problem of determining the stress state for an axisymmetric circular continuous infinite cylinder in a state of plane deformation with given incompatible deformations caused by constant the Ricci tensor in the absence of mass forces.

In this case, the Ricci tensor will be determined only by the curvature. Indeed, the strain tensor in this case has the form

$$\varepsilon^{(u)} = \begin{pmatrix} \varepsilon^{(u)}_{rr}(r) & 0 & 0 \\ 0 & \varepsilon^{(u)}_{\phi\phi}(r) & 0 \\ 0 & 0 & 0 \end{pmatrix}. \tag{40}$$

The Ricci tensor according to (26) will have the form

$$\widehat{R}^{(u)} = \begin{pmatrix} \frac{d^2}{dr^2}\varepsilon^{(u)}_{\phi\phi} + \frac{2}{r}\frac{d}{dr}\varepsilon^{(u)}_{\phi\phi} - \frac{1}{r}\frac{d}{dr}\varepsilon^{(u)}_{rr} & 0 & 0 \\ 0 & \frac{d^2}{dr^2}\varepsilon^{(u)}_{\phi\phi} + \frac{2}{r}\frac{d}{dr}\varepsilon^{(u)}_{\phi\phi} - \frac{1}{r}\frac{d}{dr}\varepsilon^{(u)}_{rr} & 0 \\ 0 & 0 & 0 \end{pmatrix}. \tag{41}$$

The scalar curvature is equal to

$$R^{(u)} = 2\left(\frac{d^2}{dr^2}\varepsilon^{(u)}_{\phi\phi} + \frac{2}{r}\frac{d}{dr}\varepsilon^{(u)}_{\phi\phi} - \frac{1}{r}\frac{d}{dr}\varepsilon^{(u)}_{rr}\right). \tag{42}$$

Thus, the components of the Ricci tensor are:

$$R^{(u)}_{rr} = R^{(u)}_{\phi\phi} = \frac{1}{2}R^{(u)}. \tag{43}$$

Beltrami-Mitchell equations can be written as follows

$$\frac{\partial^2\sigma_{rr}}{\partial r^2}+\frac{1}{r}\frac{\partial\sigma_{rr}}{\partial r}-\frac{2}{r^2}\left(\sigma_{rr}-\sigma_{\phi\phi}\right)+\frac{2\lambda+2\mu}{3\lambda+2\mu}\frac{\partial^2\sigma}{\partial r^2}+2\mu R^{(u)}_{rr}+\frac{\lambda\mu R^{(u)}}{\lambda+2\mu}=0,$$
$$\frac{\partial^2\sigma_{\phi\phi}}{\partial r^2}+\frac{1}{r}\frac{\partial\sigma_{\phi\phi}}{\partial r}+\frac{2}{r^2}\left(\sigma_{rr}-\sigma_{\phi\phi}\right)+\frac{2\lambda+2\mu}{3\lambda+2\mu}\frac{1}{r}\frac{\partial\sigma}{\partial r}+2\mu R^{(u)}_{\phi\phi}+\frac{\lambda\mu R^{(u)}}{\lambda+2\mu}=0, \quad (44)$$
$$\frac{\partial^2\sigma_{zz}}{\partial r^2}+\frac{1}{r}\frac{\partial\sigma_{zz}}{\partial r}+\frac{\lambda\mu R^{(u)}}{\lambda+2\mu}=0.$$

The system (44) must be solved in conjunction with the equilibrium equations, which in this instance will be reduced to a single equation

$$\frac{\partial\sigma_{rr}}{\partial r}+\frac{\sigma_{rr}-\sigma_{\phi\phi}}{r}=0, \quad (45)$$

Let's find a partial solution of the system of Eqs. (44)–(45). The third equation can be integrated directly, resulting in

$$\sigma_{zz}=-\frac{\lambda\mu R^{(u)}}{4(\lambda+2\mu)}r^2, \quad (46)$$

From the equilibrium Eq. (45), it can be expressed $\sigma_{\phi\phi}$ and substituted into the first Eq. (44). Then, using Eqs. (45) and (46), we obtain

$$\frac{\partial^2\sigma_{rr}}{\partial r^2}-\frac{1}{r}\frac{\partial\sigma_{rr}}{\partial r}+\frac{2\lambda+2\mu}{3\lambda+2\mu}\frac{\partial^2}{\partial r^2}\left(r\frac{\partial\sigma_{rr}}{\partial r}+2\sigma_{rr}-\frac{\lambda\mu R^{(u)}}{\lambda+2\mu}\frac{r^2}{4}\right)+\frac{2\lambda\mu+2\mu^2}{\lambda+2\mu}R^{(u)}=0, \quad (47)$$

A private solution that represents

$$\sigma_{rr}=-\frac{\mu(5\lambda+4\mu)R^{(u)}}{16(\lambda+2\mu)}r^2, \quad (48)$$

Of (44) is $\sigma_{\phi\phi}$

$$\sigma_{\phi\phi}=-\frac{3\mu(5\lambda+4\mu)R^{(u)}}{16(\lambda+2\mu)}r^2. \quad (49)$$

The general solution of a problem is the sum of a particular solution and a solution of a homogeneous problem that is known. Thus, incompatible deformations determined by scalar curvature cause additional stresses in accordance with formulas (46), (48), and (49).

References

1. Kondo, K.: On the geometrical and physical foundations of the theory of yielding. Proc. Japan Nat. Congr. Appl. Mech. **2**, 41–47 (1953)
2. Bilby, B., Bullough, R., Smith, E.: Continuous distributions of dislocations: a new application of the methods of non-Riemannian geometry. Proc. Roy. Soc. A. **231**, 263–273 (1955)
3. Kadich, A.: Edelen, D: Gauge Theory of Dislocations and Disclinations. Mir, Moscow (1987)
4. Katanaev, M.O.: Geometric theory of defects. UFN. **175**(7), 705–733 (2005)
5. Guzev, M.A., Myasnikov, V.P.: Thermomechanical model of an elastoplastic material with structural defects. MSB. **4**, 156–172 (1998)
6. Amenzade, Y.A.: Theory of elasticity. Ed. 3th textbook for universities. Higher School, Moscow (1976)
7. Rabotnov, Y.N.: Deformable Solid Body Mechanics. Nauka, Moscow (1988)
8. Dubrovin, B.A., Novikov, S.P., Fomenko, A.T.: Modern Geometry: Methods and Applications. Nauka, Moscow (1986)
9. Eliseev, V.V.: Mechanics of Elastic Bodies. Publishing House of St. Petersburg State Technical University, St. Petersburg (1999)
10. Azanov, N.P.: Saint-Venant and Bedtrami–Mitchell compatibility equations in Riemannian spaces. Tr. Geom. Sem. **19**, 9–13 (1989)
11. Nemtsev, E.A., Galimov, K.Z.: Derivation of dynamic stress compatibility conditions from the Castigliano variational principle. Res. Accord. Theory Plates Shells. **10**, 332–337 (1973)
12. Norden, A.P.: Spaces of Affine Connectivity. Nauka, Moscow (1976)
13. Pobedrya, B.E.: Lectures on Tensor Analysis. Nauka, Moscow (1974)

Optimization of Cargo Delivery Route to Nearby Cities

S. N. Masaev, S. O. Barachevskaya, and V. V. Vingert

Abstract The article is devoted to the Little method applied to solve the well-known traveling salesman problem, which consists in finding the shortest route passing through all given cities exactly once and returning to the starting point. Little's method is based on the principles of linear programming, which makes it a powerful tool for finding the optimal solution. The article describes the process of applying Little's method: first, a mathematical model of the problem is created in the format of a linear program, including constraints and objective functions. Then, feasible solutions are studied, which help to understand how to optimize the route. Data analysis is carried out step by step, which ultimately allows us to determine the optimal route that minimizes the total path length. To illustrate the methodology, an example with five cities was presented. In the course of research, an optimal solution was found that not only reduces the distance, but also significantly improves logistics, reducing transportation costs and travel time. It is analyzed that the little method is not only theoretically significant, but also practically applicable in solving complex logistical problems.

Keywords Algorithm · Intelligence · Optimization · Routes · Distances · Logistics · Processes

S. N. Masaev
Reshetnev Siberian State University of Science and Technology, Krasnoyarsk, Russian Federation

Siberian Federal University, Krasnoyarsk, Russian Federation

S. O. Barachevskaya (✉)
Reshetnev Siberian State University of Science and Technology, Krasnoyarsk, Russian Federation

V. V. Vingert
Siberian Federal University, Krasnoyarsk, Russian Federation

A. Gibadullin, S. Gordin (eds.), *Computing Technologies and Applied Mathematics*, Springer Proceedings in Mathematics & Statistics 500,
https://doi.org/10.1007/978-3-031-90546-9_10

1 Introduction

The traveling salesman problem is a classical optimization problem that consists of finding the shortest path or cycle that passes through a given set of cities and returns to the starting city, visiting each city only once. The solution to this problem has wide practical applications in various fields, from planning delivery routes to optimizing manufacturing processes. The traveling salesman method is an algorithm or mathematical approach used to solve this problem. It allows finding the optimal solution by minimizing the distance traveled or the time spent. There are many algorithms that solve the traveling salesman problem, each of which has its own advantages and disadvantages [1–3].

The Little, Murty, Sweeney, and Carell algorithms, developed in the mid-twentieth century, were the first branch-and-bound algorithms specifically designed to solve the traveling salesman problem. They initiated the development of more efficient methods that provide more accurate solutions. Branch-and-bound algorithms are still considered the most accurate methods for solving the traveling salesman problem. The Bellman–Moore–Taragan (BMT) algorithm and its modifications are used for asymmetric distance matrices, and the Held–Karp (HK) algorithm and its modifications are used for symmetric matrices.

The application of the traveling salesman method in delivery logistics allows for significant optimization of routes, reducing delivery time and minimizing transport costs. Given the specifics of the transport infrastructure, the edges of the graph used to model the problem can be adjusted to reflect real-world constraints such as one-way streets, prohibited zones, and traffic jams.

In modern conditions, the use of specialized software products based on traveling salesman methods allows logistics companies to optimize their routes, reducing delivery time, fuel consumption and the number of vehicles.

Although the traveling salesman problem is NP-complete, meaning that the time it takes to solve it increases exponentially with the number of cities, modern algorithms and technologies make it possible to solve this problem efficiently even for very large data sets [3, 4].

Advantages of using the traveling salesman method in logistics:

Reduced delivery costs: route optimization allows you to reduce mileage, fuel consumption and driver wage costs.

Increased delivery efficiency: Reducing delivery times allows you to process more orders, improving response speed and increasing customer satisfaction.

Reduced CO_2 emissions: Optimising routes reduces the number of vehicles on the road, which in turn reduces greenhouse gas emissions.

Thus, the traveling salesman method is a powerful tool for optimizing logistics processes, which allows enterprises to increase their efficiency, reduce costs and reduce their impact on the environment [5].

2 Method

Let us consider an exact algorithm based on the detailing of general constructions for the implementation of an elementary operation for the traveling salesman problem. The statement of the traveling salesman problem is as follows. There are cities n, designated by numbers $i \in N = (1, \ldots, n)$, the distances between all these cities are given by real numbers $c_{ij} \geq 0$. When leaving a given city, you need to go around the remaining cities $(n - 1)$ at least once and then return to the given one so that the total distance traveled is minimal.

Let us present a formal scheme of Little's algorithm. We denote the sum of constants as h^s, where s is the ordinal number of the reduction.

1. Let us assume $s = 1$ that we will carry out the reduction of the matrix C by rows and columns, passing to the reduced matrix C^s.
2. We calculate the sum of the reducing constants h^s—the estimate for the set X—and denote it as $\omega(X)$.
3. We select candidates for inclusion in the set Y, i.e. all those (i,j), $i,j = \overline{1,n}$, for which $c_{ij}^s = 0$.
4. We are calculating

$$\theta(i,j) = c_{ik} + c_{lj} \tag{1}$$

for the selected candidates and select $\theta(k,l) = \max \theta(i,j)$ for all i, j for which $c_{ij}^s = 0$. The pair (k,l) is included in the set Y and is forbidden for $\overline{Y}$.

5. Estimation for a set:

$$\omega(\overline{Y}) = \omega(X) + \theta(k,l) \tag{2}$$

6. To exclude the possibility of forming closed subcycles of shorter length than n, we assume that $c_{lk} = \infty$ we will exclude the k row and l column from further consideration.
7. If the truncated matrix has dimension 2 × 2, it means that it contains only two pairs of cities that are closing in the route without subcycles of shorter length than n. We check whether the truncated matrix has such dimension. If it does, go to step 9, if not, go to step 8.
8. If the truncated matrix is not reduced, then we will carry out its reduction by calculating h^{s+1}, after which

$$\omega(Y) = \omega(X) + h^{s+1} \tag{3}$$

we accept

$$s = s + 1 \tag{4}$$

and go to step 3.

9. We will check the optimality conditions: if the evaluation of the cycle obtained after branching does not exceed the evaluations of all other possible routes, then this cycle will be optimal. If there is at least one route with a lower evaluation, then we remember the current cycle. Then we continue the branching process based on the route with the lower evaluation.

3 Choosing the Delivery Method

Modern logistics offers a variety of cargo delivery options, each with its own unique advantages and disadvantages. In this comparative analysis, we will take a closer look at the three most common types of transportation: river, rail, and road. These types of delivery are widely used in various industries, and understanding their features will help you make a more informed choice depending on your needs and business specifics.

In the field of logistics, the choice of transport for the transportation of goods plays a key role, especially when it comes to short-distance transportation. Let's look at the three main methods of delivery—road, rail and river transport—using the example of transporting cargo weighing 1 kg along a route whose cities are located on the same river. We will designate them with the letters A, B and C. The route A—B—C is 92 km long.

1. Automobile transport. According to online calculators, the cost of transporting 1 kg of cargo by car is approximately 300 rubles. Automobile transport offers high flexibility and includes the possibility of door-to-door delivery. This is especially convenient when a sea of subsequent transfers or additional processes are not required, and delivery can be completed quickly. However, it should be noted that the cost of automobile delivery can increase significantly if it is necessary to transport large volumes or if the traffic situation worsens, which is due to fluctuations in fuel and transportation prices.
2. Rail transport. Prices for rail transportation on this route are only about 3 kopecks per kilogram. This makes rail transport the most attractive option in terms of savings when transporting large volumes of cargo. However, it is necessary to remember some limitations. Rail transportation requires a developed infrastructure—stations, tracks and other logistics base. It should also be taken into account that train schedules are less flexible than road routes, which may require additional waiting time.
3. River transport. As for river transport, the cost of delivery of 1 kg of cargo is about 3600 rubles. Despite the tempting opportunity to transport large volumes, river transport has its tangible disadvantages. Firstly, high prices. Secondly, it is subject to seasonal changes, which can limit the possibility of transportation in certain periods of the year. In addition, a transfer to another type of transport is required to deliver to the final destination, which can increase the overall travel time.

Thus, each mode of transport has its advantages and disadvantages, and the choice of delivery method depends on many factors, including budget, required delivery speed and availability of infrastructure. For the transportation of a 1 kg load on the considered route, rail transport is the most cost-effective option due to its low cost. However, if speed and convenience are important, road transport may be the best solution, despite the higher costs. River transport is not very competitive in price on this route and has many logistical limitations. It is important to carefully evaluate all options to choose the most suitable for your needs.

4 The Practical Part

The traveling salesman problem, where one must find the shortest route that passes through all given cities and returns to the starting point, is a classic optimization problem. Little's method is one of the most efficient algorithms for solving it, which is usually implemented using software.

Let's imagine that we need to deliver goods to several cities. For example, cities located next to each other were taken. Let's denote them with numbers.

Approximate route order:

1. 1 – 2 – 3 – 4 – 5 – 1.
   If we accept this order and return to city 1, the car will travel 303 km.
   However, this route is not necessarily optimal. The Little method allows you to find the optimal solution, that is, the shortest possible route. This requires:
2. Define a distance matrix: create a table where each cell corresponds to the distance between two cities.
3. Apply Little's algorithm: the algorithm sequentially goes through all possible route combinations, using the principle of dynamic programming to cut off unpromising options.
4. Find the optimal route: Little's algorithm will eventually build the shortest route that goes through all the cities and returns to city 1.

Table 1 contains a sequence of cities, with a number assigned to each city and the coordinates of these cities.

Table 1 Coordinates of cities

№	*X*	*Y*
1	56.005436	92.831898
2	55.958354	92.379699
3	56.18123	92.507142
4	56.110042	93.352096
5	56.21322	93.488932

Little's method, also known as the branch and bound method, is based on the fact that the matrix of distances between all pairs of points can be used efficiently to solve the traveling salesman problem.

Initially, the matrix is filled with the distance values between all pairs of points given in the problem conditions. This means that each cell of the matrix contains information about the distance between two specific points of the route.

The search for the optimal route then takes place. Little's method uses the branch-and-bound principle to systematically explore all possible routes. The algorithm works as follows:

1. Building a decision tree. The root of the tree is the starting point of the route. Each tree node represents a city, and each branch is a possible transition between cities.
2. Limiting the boundaries. For each tree node, the lower bound of the cost of the route passing through this node is calculated. The lower bound is defined as the sum of the distances between the cities already visited.
3. Branching. The algorithm selects the node with the smallest lower bound and creates new nodes corresponding to all unvisited cities.
4. Pruning. If the lower bound of a node exceeds the cost of the best route found, this node is discarded.
5. Repeating steps 2–4 until all optimal routes are found.

Little's method allows one to effectively eliminate many unpromising routes, thereby significantly accelerating the process of finding a solution.

Figure 1 shows possible routes between cities.

As a result, the program finds the optimal path for the traveling salesman problem presented in Table 2, which minimizes the path length and passes through all the given points.

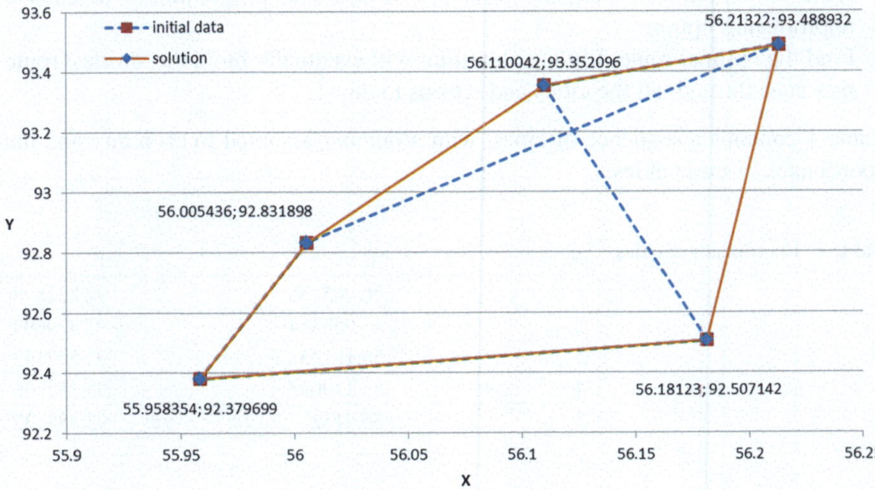

Fig. 1 Possible routes

Table 2 The optimal route

Subsequence	Distance
1	
4	0.5
5	0.2
3	1.0
2	0.3
Length of the path	2.4

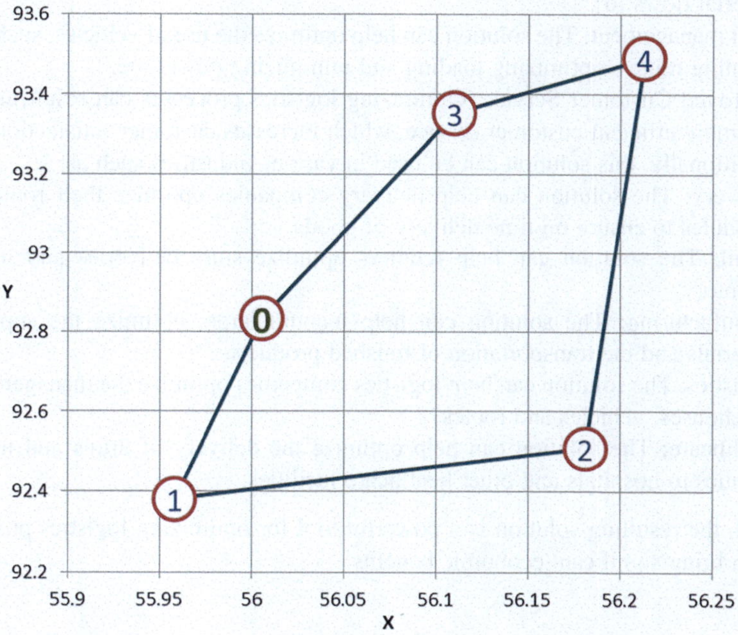

Fig. 2 The shortest route

Table 3 Percentage improvement of the target route

Found	Given	Improved function	
2.40	2.42	0.02	1%

Based on the program output, we understand that the most optimal route is: 1 – 5 – 3 – 4 – 2 – 1 with a distance of 240 km, which is shown in Fig. 2.

The percentage improvement of the target route is shown in Table 3.

The resulting solution can be used to optimize logistics processes, plan delivery routes and other tasks related to route optimization.

In particular, this solution can be used for:

- Delivery route optimization: The solution can help find the most efficient route for delivering goods, taking into account factors such as distance, travel time, capacity constraints and road conditions.
- Delivery schedule planning: The solution can help plan optimal delivery schedules, taking into account factors such as operating hours, order volume and vehicle availability.
- Warehouse management: The solution can help optimize the allocation of resources in the warehouse, such as optimizing product placement and managing material flows [6].
- Fleet management: The solution can help optimize the use of vehicles, such as by planning routes, optimizing loading and minimizing downtime.
- Improved Customer Service: Optimizing logistics processes can result in faster and more efficient customer service, which increases customer satisfaction.
- Additionally, this solution can be used in various industries such as:
- Delivery: The solution can help delivery companies optimize their routes and schedules to ensure on-time delivery of goods.
- Retail: The solution can help retailers optimize store deliveries and manage inventory.
- Manufacturing: The solution can help manufacturers optimize the supply of materials and the transportation of finished products.
- Logistics: The solution can help logistics companies optimize the management of warehouses, vehicles and routes.
- Healthcare: The solution can help optimize the delivery of drugs and medical supplies to hospitals and other healthcare facilities.

Overall, the resulting solution is a powerful tool for optimizing logistics processes and can bring significant economic benefits.

5 Overview of Similar Tasks

In his work, V. O. Boroznov presented a comprehensive research of the traveling salesman problem, analyzing the effectiveness of various methods for solving it.

For each problem size N, 33 trials were conducted, which provided a confidence level of 0.75 and an acceptable error of 0.1.

The essence of the experiment was that for each dimension N a distance matrix was randomly generated, which was then used to solve the problem using four different methods: a genetic algorithm, a BV method, an ACS-Q method, and another non-classical genetic algorithm using Greedy Crossover. The results of the research showed that the genetic algorithm as a whole turned out to be the most accurate method for solving the traveling salesman problem. It should be noted that the use of a non-classical GA with Greedy Crossover can further reduce the error and increase the accuracy of the solution.

The BV method and the ACS-Q method demonstrated lower efficiency compared to the genetic algorithm. This is explained by the fact that the distance matrix used values that were close or equal to each other. Under such conditions, as experiments have shown, these methods do not cope with the task as well as the classical genetic algorithm.

Thus, the research conducted by V. O. Boroznov confirms the high efficiency of genetic algorithms for solving the traveling salesman problem, especially in cases where the distance matrix contains close values.

It is also important to note that the use of non-classical genetic algorithms, for example, with Greedy Crossover, allows to increase the accuracy of the solution and expand the possibilities of solving the traveling salesman problem in various conditions.

This work makes a valuable contribution to the development of methods for solving the traveling salesman problem, especially in the context of using genetic algorithms [7].

This article by Jalynne Figueroa, Becky Miller, and Mai Okimoto focuses on the unique case of Wal-Mart and Kmart's strategies. Wal-Mart, being one of the largest retailers in the world, actively uses the Traveling Salesperson Problem (TSP) to optimize delivery routes.

How the TSP works at Wal-Mart:

- Identifying points: Wal-Mart identifies all of its stores, warehouses, and suppliers as "nodes" in a route.
- Calculating distances: Using GPS and mapping services, Wal-Mart calculates the distance between each "node."
- Build an optimal route: TSP's specialized algorithms, based on various mathematical approaches (e.g., genetic algorithms, simulated annealing), build an optimal route that minimizes the total distance or delivery time.
- Route Optimization: In addition to TSP algorithms, Wal-Mart also uses various methods for route optimization, such as order bundling, route merging, and delivery time optimization.

Benefits of using TSP for Wal-Mart:

- Reduced logistics costs: Optimal routes reduce fuel costs, driver time, and vehicle wear and tear.
- Improved delivery efficiency: Optimized routes allow more deliveries to be made in less time, resulting in faster and more reliable delivery of goods.
- Increased customer satisfaction: Fast and timely delivery of goods increases customer loyalty and enhances Wal-Mart's reputation.

In the work of A. V. Martynov and V. M. Kureichik, an effective hybrid algorithm is proposed for solving the classical traveling salesman problem, combining the advantages of the ant algorithm (AA) and the genetic algorithm (GA). The ant algorithm, being a metaheuristic, imitates the behavior of an ant colony when searching for the optimal path to a food source. The algorithm uses the concept of pheromone trails: ants leave pheromones along the path they have traveled, which allows subsequent ants to choose shorter and more effective routes. In this work, AA is used to find the shortest route in a graph, supporting the most adapted agents that pass on genetic information to subsequent generations.

The genetic algorithm, as a metaheuristic, is based on the processes of natural selection and evolution. GA uses a population of chromosomes, each of which represents a solution to the problem. Within the framework of the algorithm, genetic recombination (crossing over) and random mutations occur, which allows for the formation of new, potentially higher-quality solutions.

The hybrid algorithm combines these two metaheuristics, integrating GA into the MA process. The hybrid algorithm is based on an approach where MA is used to find initial approximations, which are then optimized using GA. The key element of the hybrid algorithm is the developed method for evaluating the routes obtained by MA. The evaluation allows identifying the most promising routes, which are then used as input data for GA. The paper also proposes a heuristic aimed at optimizing the use of routes obtained by MA in GA.

The experimental results obtained by the authors confirm the effectiveness of the proposed hybrid algorithm. The hybrid algorithm demonstrates a higher quality of solutions compared to the conventional MA, finding optimal routes in the same time. In addition, the average quality of solutions for multiple runs of the application is also higher for the hybrid algorithm [8].

6 The Use of the Traveling Salesman's Task in the Special Economic Zones of the Russian Federation for the Transportation of Goods by River Transport

The use of TSP in special economic zones (SEZ) for the transportation of goods by river transport can be very relevant:

- Optimization of logistics chains: As a rule, SEZs concentrate various enterprises that require an effective cargo delivery system.
- Reduction of transport costs: The optimal route allows minimizing the mileage of vessels, which reduces fuel costs and delivery time.
- Increase in throughput: properly planned routes contribute to more efficient use of vessels and river transport infrastructure.
- Increase in work efficiency: Minimizing idle runs allows increasing labor productivity and optimizing the work of transport companies.
- However, the application of TSP in this context has its own peculiarities:
- Complexity of conditions: River routes can be limited by various factors, such as water levels, the presence of locks, cargo capacity restrictions, the presence of bridges.
- Dynamic changes: River conditions can change, which requires adaptation of routes and operational planning.

- Need to consider the specifics of cargo: Some cargoes require special transportation conditions, which can affect the choice of the optimal route.

How to use TSP in SEZ for river transport:

- Data collection: It is necessary to obtain information about the geographical location of enterprises, river routes, transportation restrictions, time and cost factors.
- Use of specialized software: there are various algorithms and programs for solving the traveling salesman problem, including machine learning methods and neural networks.
- Adaptation to the specifics of river transport: It is necessary to take into account the specifics of river transport, such as the presence of locks, restrictions on cargo capacity, seasonal changes.
- Monitoring and adaptation: Regular monitoring and analysis of the obtained data will optimize routes and improve the efficiency of logistics processes.

The use of the traveling salesman problem in special economic zones of the Russian Federation for the transportation of goods by river transport can significantly improve the efficiency of logistics chains, reduce transportation costs and increase the capacity of river transport. However, it is necessary to take into account the specifics of river transport and use appropriate software solutions to solve the TSP [9, 10].

7 Conclusion

In conclusion, this research demonstrates the high efficiency of Little's method in solving the classical traveling salesman problem. The method, based on the powerful concept of linear programming, allows finding the optimal route with high accuracy, passing through all the given cities only once and returning to the starting point. The article describes in detail the process of applying Little's method, which includes several key stages:

- Building a model of the problem: The first step is to translate the traveling salesman problem into a mathematical model using the language of linear programming. This involves defining the variables, constraints, and objective function that describe all aspects of the problem.
- Finding feasible solutions: Once the model has been built, it is necessary to find a set of feasible solutions, i.e. routes that satisfy all the constraints of the problem. This can be done using various methods, such as the simplex method or the branch and bound method.
- Determining the optimal route: The final step is to select from among all the feasible solutions the route that minimizes the objective function, in this case, the total route length. This is achieved using optimization methods, such as the simplex method, the branch and bound method, or other specialized algorithms.

The practical application of Little's method to the five-city traveling salesman problem presented in the paper clearly demonstrates its ability to reduce the total route length, thereby optimizing logistics and minimizing costs. The results of the research highlight the value of Little's method for solving a wide range of route optimization problems arising in various fields, such as transportation, logistics, delivery, planning, etc.

Moreover, it is important to note that Little's method is flexible and adaptable. It can be easily modified to take into account additional constraints such as delivery time constraints, load capacity constraints, road capacity constraints, etc. Overall, Little's method is a powerful tool for solving route optimization problems and can be effectively applied in various fields of activity. The results of this research confirm its effectiveness and reliability, making it a valuable tool for optimizing logistics processes and minimizing costs [11].

References

1. Krotov, V.F., Lagosha, B.A., Lobanov, S.M., Danilina, N.I., Sergeev, S.I.: Fundamentals of Optimal Control Theory. Higher School, Moscow (1990)
2. Little, J.D.C., Murty, K.G., Sweeney, D.W., Karel, C.: An algorithm for the traveling salesman problem. Oper. Res. **11**, 972–989 (1963)
3. Shatalova, N.V.: Russian Transport: Problems and Prospects—2019. Institute of Transport Problems of the Russian Academy of Sciences, Saint-Petersburg (2019)
4. The traveling salesman's task—the method of branches and boundaries. Galyautdinov—economics teacher's website, https://galyautdinov.ru/post/zadacha-kommivoyazhera. Accessed 27 Nov 2023
5. Mudrov, V.I.: The Traveling Salesman Problem. Znanie, Moscow (1969)
6. Dorrer, M.: Model of optimal organization maturity management under conditions of interference and uncertainty. J. Phys. Conf. Ser. **1399**, 33086 (2019)
7. Boroznov, V.O.: Investigation of the solution of the traveling salesman problem. Bull. Astrakhan State Techn. Univ. **2**, 147–151 (2009)
8. Martynov, A.V., Kureichik, V.M.: Hybrid algorithm for solving the traveling salesman problem. Izvestiya Yuzhnogo Federal'nogo Universiteta. **4**(165), 36–44 (2015)
9. Takashima, Y., Nakamura, Y.: Theoretical and experimental analysis of traveling salesman walk problem. IEEE Asia Pacific Conference on Circuit and Systems (APCCAS), pp. 241–244 (2021)
10. Ding, C., Cheng, Y., He, M.: Two-level genetic algorithm for clustered traveling salesman problem with application in large-scale TSPs. Tsinghua Sci. Technol. **4**, 459–465 (2007)
11. Hu, B., Raidl, G. R.: Solving the Railway Traveling Salesman Problem via a Transformation into the Classical Traveling Salesman Problem. Eighth International Conference on Hybrid Intelligent Systems (2008)

Refinement of Variable Order Fractional Derivative of Gerasimov-Caputo Type by Multidimensional Levenberg-Marquardt Optimization Method

Dmitriy Tverdyi

Abstract When solving many fundamental and applied problems related to the study of properties of dynamical systems, situations often arise when the parameters of the model are not exactly defined. One can try to solve such a problem by searching the parameters, taking into account the understanding of the nature of the modelling process. On the other hand, it is possible to solve inverse problems—a common type of problems in many scientific fields, where it is necessary to determine the values of the model parameters on the basis of experimental data, but it is impossible to make direct measurements of the parameters. The paper considers an inhomogeneous fractional equation with a Gerasimov-Caputo type operator of variable $\alpha(t)$ order. The possibility of recovering the form of the function $\alpha(t)$ and refining its values on the basis of test generated data is investigated. The direct problem is defined as a Cauchy problem for a fractional equation solved numerically. The solution of the inverse problem is reduced to the minimization of the inviscid functional, and the minimization problem is solved using the iterative Levenberg-Marquardt unconditional optimization method. Thus, the problem of determining the optimal form $\alpha(t)$ of a fractional equation is reduced to 4 parameters that control the course of the solution of the inverse problem and do not depend on the physical meaning of the original problem in specific models. On test examples it was shown that the Levenberg-Marquardt method can indeed be used for unconditional optimization to determine type of function $\alpha(t)$ and its optimal values.

Keywords Inverse problems · Non-conditional optimization · Levenberg-Marquardt algorithm · Fractional derivatives · Gerasimov-Caputo · Memory effect · Non-locality · Implicit finite-difference schemes

D. Tverdyi (✉)
Institute of Cosmophysical Research and Radio Wave Propagation FEB RAS, Paratunka, Russia

A. Gibadullin, S. Gordin (eds.), *Computing Technologies and Applied Mathematics*, Springer Proceedings in Mathematics & Statistics 500,
https://doi.org/10.1007/978-3-031-90546-9_11

1 Introduction

Integral-differential calculus has been actively developing for more than 300 years, and during this time it has proved to be a reliable and powerful tool for solving theoretical and applied problems. Throughout the history of this branch of mathematical science, researchers have been fascinated by the question of the possibility of generalizing the concepts of integral and derivative from integer to real order [1]. Since the time of G. W. Leibniz and G. F. Lopital (1965), various definitions of the concept of fractional differentiation operator have been developed: Riemann-Liouville, Weyl, Grunwald-Letnikov and many others. Also, applications of fractional derivatives and integrals are increasingly being found in areas where the integer definition was insufficient. In quantum physics [2], viscoelasticity [3, 4] and solid mechanics [5], anomalous diffusion processes [6], fractal oscillators [7], etc.

In our study, fractional derivatives (FD) are interesting because they are useful for illustrating the memory and hereditary properties of many dynamical processes [8]. This effect is also called hereditary, and one of the first scientists to work on deriving this theory was the Italian mathematician and physicist V. Volterra [9]. Hereditary and its relation to fractional derivatives can be shown as follows. Consider the following integro-differential equation (IDE):

$$\int_{-\infty}^{t} K(t-\sigma)\dot{u}(\sigma)d\sigma = F(u(t),t), \tag{1}$$

where, according to V. Voltera [9] principles of heredity, memory takes into account all possible prehistory. Since in practice we do not have data on the entire history of the process, we should consider the integral on the sub-interval $(0, t)$ where, $t \in [0, T]$ is the current simulation time, $T > 0$ is the total simulation time. The main point is that choosing the form of the kernel $K(t-\sigma)$ as the degree function $K(t-\sigma) = \frac{(t-\sigma)^{-\alpha}}{\Gamma(1-\alpha)}$ allows us to pass to the fractional calculus operators [7, 10].

Turning to the FD formalism when dealing with such IDE, in our study we will consider the integral in the left-hand side of (1) in terms of the Gerasimov-Caputo fractional differentiation operator [11, 12] of constant $0 < \alpha < 1$ order:

$$\int_0^t \left(\frac{(t-\sigma)^{-\alpha}}{\Gamma(1-\alpha)}\right)\dot{u}(\sigma)d\sigma \equiv \frac{1}{\Gamma(1-\alpha)}\int_0^t \frac{\dot{u}(\sigma)}{(t-\sigma)^{\alpha}}d\sigma \equiv \partial_{0,t}^{\alpha}u(\sigma), \tag{2}$$

where, $\Gamma(\cdot)$ is the Euler gamma function; $\dot{u} = du/dt$.

The memory effect is represented by the dependence of the current state of the dynamical system on the previous states, i.e. non-locality. The intensity of this effect will be determined by the value of the degree of FD. Classically, this certain α value is non-integer and constant [13].

In science and engineering, memory properties are common in complex systems. However, there are generalizations of fractional derivatives to the case of time-variable non-locality $\alpha(t)$ and other functional dependencies [14]. Variable order fractional derivatives (VOFD) find their application in the theory and practice of physical and mathematical sciences. For example, the diffusion equations [15] or the dynamic viscoelastic oscillator model [16].

In [17] generalization FD of the form (2) to VOFD at $0 < \alpha(t) < 1$ studied in detail:

$$\frac{1}{\Gamma(1-\alpha(t))}\int_0^t \frac{\dot{u}(\sigma)}{(t-\sigma)^{\alpha(t)}}d\sigma \equiv \partial_{0,t}^{\alpha(t)}u(\sigma), \tag{3}$$

where, $\alpha(t) \in C^1(0, 1)$ is a continuous-differentiable function.

In such a case, the Eq. (1) in the domain $\Omega = \{(t) : 0 \leq t \leq T\}$ can be represented as a fractional equation (FDE) of the form:

$$\partial_{0,t}^{\alpha(t)}u(\sigma) = F(u(t), t), \tag{4}$$

where, $u(t) \in C^2(0, T)$ a twice continuously differentiable solution function; $F(u(t), t)$ a functional defining the form of the FDE.

The series of works of the author of this research, including publications [18, 19], is devoted to the study of issues related to mathematical modeling of volumetric activity of radon gas (RVA) in accumulation chambers with gas-discharge sensors. Accumulation chambers are located in observation points, which are located above the zones of dynamic influence of faults in the Earth's crust. These zones have increased permeability, which favors the flow of subsurface gases into the atmosphere [20]. It is assumed that the parameter $\alpha(t)$ describes the fractality of the geosphere and is related to its characteristics such as porosity, permeability and fracturing, which directly affect the intensity of the process. As a result, a number of mathematical models based on nonlinear FDE of the form (4) have been developed. In [18] it is shown that nonlinear models with FD of the form (2) of constant order α can well describe the accumulative modes of RVA. In [18] it is shown that generalization to VOFD of the form (3) to the case of $\alpha(t)$ also allows describing some anomalous RVA modes. Continuous monitoring of RVA variations in order to detect anomalies in its values is one of the effective methods to investigate the stress-strain state of the geosphere. RVA is considered to be an informative and operative precursor of seismic events, which determines the relevance of such studies.

When solving many fundamental and applied problems related to the study of certain properties of dynamical systems, situations often arise when the parameters of the model are not precisely determined. However, if experimental data or information on the ranges of values of the model parameters are known, it is possible to select various parameters of such models empirically, taking into account the available understanding of the nature of the modeled process. In the [18, 19] articles, the model parameters are refined by enumerating values and comparing time series: modeling results and experimental data representing the process. This process continues until the modelling results begin to qualitatively approximate the data, that is, until high values of correlation and R^2 determination [21] are reached. This approach is time consuming, which inevitably leads us to ideas of different ways to automate the selection of optimal parameters.

For example, in the research [22, 23] for fractional equations, numerical methods for obtaining interval estimates of solutions from known interval values of parameters for anomalous diffusion problems are applied.

An inverse problem can be solved to obtain a solution to the problem of finding the optimal parameters. Such problems are often ill-posed [24], i.e., the solution may not exist, the solution may not be unique, or the solution is unstable [25]. However, the inverse problem is a common type of problem in many scientific fields [26], where it is necessary to determine the values of model parameters from observed data [27]. Despite the fact that this branch of mathematics began to be actively developed only in the XIX-XX century, it can rightly be considered one of the most important. The need for such an approach often arises in geological data [29], in geophysics and seismology [30], in computed tomography [31, 32] and many others. This issue is especially acute for geophysics, as it is impossible to make direct measurements deep underground [27].

2 Direct Problem Statement

Consider the case when the functional $F(u(t), t) = 1$. Then (4) is defined as an inhomogeneous FDE for which in the domain $\Omega = \{(t) : 0 \leq t \leq T\}$ we consider a Cauchy problem of the form:

$$\partial_{0,t}^{\alpha(t)} u(\sigma) = 1, \quad u(0) = u_0, \tag{5}$$

where, $\partial_{0,t}^{\alpha(t)} u(\sigma)$—FD of variable order $0 < \alpha(t) < 1$ of the form (3); $u(t)$—solution function; $\alpha(t)$—degree of the fractional derivative of (4); $u(t) \in C^2(0, T) = U$—the class of twice continuously differentiable functions; $\alpha(t) \in C(0, 1) = A$—class of continuous-differentiable functions; u_0—some constant, the initial condition of the Cauchy problem.

Definition 1 The direct problem is a Cauchy (5) problem consisting of finding $u(t) \in U$ given a known $\alpha(t) \in A$.

To solve the direct problem (5), we use the previously developed and theoretically justified non-local implicit finite-difference scheme (IFDS), which has been tested on a number of test problems [17]. We also note that (5) can be solved using the explicit finite-difference scheme (EFDS) presented in [17, 28]. Let a uniform grid domain $\widehat{\Omega}$ be given, in which classes of grid functions $\widehat{U}$ and $\widehat{A}$ with a sampling step τ are defined:

$$\tau = T/N, \widehat{\Omega} = (t_i = i\tau) : 0 \leq i < N, \widehat{U} \in \widehat{\Omega}, \widehat{A} \in \widehat{\Omega},$$
$$u(t) = u(t_i) = u_i, \alpha(t) = \alpha(t_i) = \alpha_i, 0 < \alpha_i < 1, \tag{6}$$

The in-homogeneous FDE (5) can be approximated in the domain (6) by the IFDS scheme as follows:

$$A_i \sum_{j=0}^{i-1} w_j^i \left(u_{i-j} - u_{i-j-1}\right) - 1 = 0, u_0 = C,$$
$$A_i = \frac{\tau^{-\alpha_i}}{\Gamma(2 - \alpha_i)}, w_j^i = (j+1)^{1-\alpha_i} - j^{1-\alpha_i}, 1 \leq i < N, \tag{7}$$

where C—is a known constant.

Definition 2 The difference direct problem (7) is to find the grid $u(t_i) \in \widehat{U}$ in the domain (6) for known values of the grid function $\alpha(t_i) \in \widehat{A}$.

The IFDS scheme (7) can be solved by Newton's method or by a modified Newton's method. In [17], a number of theorems on the convergence and stability of IFDS for FDEs of the more general form (4) as well as fractional derivative approximation (3) were formulated and proved. More details on the numerical method, see in [17, 33].

Remark 1 The difference direct problem (7) based on the IFDS scheme and solved by Newton's method, according to the [17], is unconditionally stable and converges with order $O\left(\tau^{2 - max\, 1_t(\alpha(t))}\right)$.

3 Inverse Problem Statement

Let the values of the function $\alpha(t) \in A$ (or its grid analogue $\alpha(t_i) \in \widehat{A}$) be unknown. However, additional information (experimental data) is known about the solution of the difference direct Cauchy problem (7) in the domain $\widehat{\Omega}$:

$$u(t_i) = \theta(t_i), \tag{8}$$

Let $\alpha(t) \in A$ be a function of some known class, and its form is clearly defined by some set of parameters. Then, according to the book [34] by Tikhonov A.N., Samarskii A.A., the solution of the inverse problem is reduced to the search for the values of these parameters. Following this approach, let us define the set of vectors $\overrightarrow{\mathbb{X}}$ as the space of solutions to the inverse problem (5). The vector $\overrightarrow{X} =$

$[X_0, \ldots, X_{K-1}]$ a set of unknown parameters characterizing the form of the function $\alpha(t)$, where $\vec{X} \in \vec{\mathbb{X}}$, $\vec{\mathbb{X}} \subset \mathbb{R}^K$, K is the number of components.

Definition 3 The inverse problem for (5) is defined as the recovery (definition) of the function $\alpha(t) = \alpha\left(\vec{X}\right) \in A$:

$$\partial_{0,t}^{\alpha\left(\vec{X}\right)} u(\sigma) = 1, u(0) = \theta_0, \tag{9}$$

Definition 4 Therefore, the difference inverse problem for (7) is defined as the restoration of the grid function $\alpha(t_i) = \alpha\left(\vec{X}\right) = \alpha(X_0, \ldots, X_{K-1}) \in \hat{A}$, from the known (8) additional information:

$$A_i \sum_{j=0}^{i-1} w_j^i (u_{i-j} - u_{i-j-1}) - 1 = 0, u_0 = \theta_0, u_i = \theta_i,$$
$$A_i = \frac{\tau^{-\left(\alpha\left(\vec{X}\right)\right)}}{\Gamma\left(2 - \left(\alpha\left(\vec{X}\right)\right)\right)}, w_j^i = (j+1)^{1-\left(\alpha\left(\vec{X}\right)\right)} - j^{1-\left(\alpha\left(\vec{X}\right)\right)}, 1 \le i < N, \tag{10}$$

Let $\vec{\theta} = [\theta_0, \ldots, \theta_{N-1}]$ the vector of experimental data according to (8), and $\omega(t_i) \in \hat{U}$, $\omega(t_i) = \omega\left(\vec{X}\right) = [\omega_0, \ldots, \omega_{N-1}]$ is the vector of model data, i.e., the solution of the difference direct problem (7) obtained with respect to the approximation $\alpha\left(\vec{X}\right)$ at some given values of $\vec{X}$.

Let $\vec{\eta} = \eta\left(\vec{X}\right) = [\eta_0, \ldots, \eta_{N-1}]$ be a bias vector of dimension $N > K$ such that $\mathbb{R}^K \to \mathbb{R}^N$, i.e., a real subset of $\mathbb{R}$ that depends on the solution $\omega\left(\vec{X}\right)$ of dimension N and hence depends on $\alpha\left(\vec{X}\right)$ of K parameters of $\vec{X}$:

$$\vec{\eta} = \vec{\theta} - \omega\left(\vec{X}\right), \tag{11}$$

Then, in terms of unconditional optimization theory, according to [35], the solution of the difference inverse problem (10) reduces to the minimization of the Ψ of bias functional:

$$\min_{\vec{X} \in \mathbb{R}^K} \Psi\left(\vec{X}\right) = M\left(\eta\left(\vec{X}\right)\right), \eta : \mathbb{R}^K \to \mathbb{R}^N, M : \mathbb{R}^N \to \mathbb{R},$$
$$M\left(\eta\left(\vec{X}\right)\right) = \frac{1}{2} \vec{\eta}^T \vec{\eta} = \frac{1}{2} \sum_{i=0}^{N-1} \eta_i^2 = \frac{1}{2} \sum_{i=0}^{N-1} (\theta_i - \omega_i)^2, \tag{12}$$

Remark 2 Since the continuous inverse problem is discretized, it often happens that exactly the stability condition obtained by numerical method is broken, defining the difference inverse problem (10) as an ill-posed [36]. However, in the considered case, the unconditionally stable IFDS algorithm [17] is used to solve the direct problem (5) in the discrete formulation (7).

4 Method of Minimizing the Target Function

To solve such a minimization problem (12), we turn to mathematical optimization methods [35, 37]. There is a class of methods called iterative descent or gradient descent methods [38, 39], but they have an unpleasant peculiarity. In practice, the values of $\vec{\eta}$ may have local minimums in addition to the desired global minimum. Depending on the choice of the initial approximation $\overrightarrow{X^0}$, the method may converge just to a local minimum of the functional $\Psi\left(\vec{X}\right)$ rather than a global minimum.

As an alternative to iterative descent methods, there is a class of Newtonian (quasi-Newtonian) methods that are among the most efficient in practice [35]. We can use Newton's unconditional optimization method, but it requires that the functional $\Psi\left(\vec{X}\right) \in C^2\left(G \subset \mathbb{R}^K\right)$ twice continuous-differentiable in an open convex set G. More details in [35]. This condition is required to compute the Hessian (Hesse matrix) of the form:

$$H = \nabla^2 \Psi\left(\vec{X}\right) = \nabla^2 \Psi\left(\vec{X}\right)_{i,k} = \frac{\partial^2 \Psi\left(\vec{X}\right)}{\partial X_i \partial X_k}, \quad i \geq 0, \quad k \leq K-1,$$

Newton's unconditional optimization method is implemented as an iterative procedure:

$$\overrightarrow{\Delta X} = \left(-H^{-1}\right) \times \left(\nabla \Psi\left(\vec{X}\right)\right),$$

where, $\nabla \Psi\left(\vec{X}\right)$—the gradient of Ψ in $\vec{X}$; approximation $\vec{X}$—the recovered values computed at the current iteration of the procedure; $\overrightarrow{\Delta X} = \left[\Delta X_0 \ldots \Delta X_{K-1}\right]$—the optimal increments $\vec{X}$ for the next iteration.

Remark 3 An important characteristic of numerical methods of optimization is the convergence rate, which characterizes the efficiency of the method used to solve the inverse problem. Therefore, it is desirable to use the method with the highest convergence rate.

Therefore, we will use a modification of Newton's method, called the Levenberg-Marquardt iterative method of unconditional optimization [40–42], which allows us to get rid of the condition for the existence of second derivatives. Assuming that at least $\Psi\left(\vec{X}\right) \in C^1\left(G \subset \mathbb{R}^K\right)$ is satisfied, and $\nabla \Psi\left(\vec{X}\right) = J^T \times \vec{\eta}$, and the Hesse matrix of dimension $K \times K$ is calculated as follows:

$$H = J^T \times J + \gamma E,$$

then the Levenberg-Marquardt method is realized as the following iterative procedure for the system of linear algebraic equations:

$$\overrightarrow{\Delta X} = \left(-\left(J^T \times J + \gamma E\right)^{-1} \right) \times \left(J^T \times \vec{\eta} \right), \tag{13}$$

where,

- E—is a unitary matrix of dimension $K \times K$;
- $\vec{\eta} = \overrightarrow{\eta^{(n)}} = \eta\left(\overrightarrow{X^{(n)}}\right)$—bias vector;
- $J = J\left(\vec{X}\right) = J\left(\overrightarrow{X^{(n)}}\right)$—is a Jacobi matrix of dimension $N \times K$ whose elements have the form: $J_{i,k} = \frac{\partial \eta_i^{(n)}}{\partial X_k^{(n)}}, i = 0..N-1, k = 0..K-1.$

and the derivatives in $J_{i,\ k}$ are approximated by a finite difference of the form:

$$J_{i,k} = \frac{\eta_i^{(n)\delta} - \eta_i^{(n)}}{\delta X_k},$$

where,

- $\overrightarrow{\delta X} = [\delta X_0 \ldots \delta X_{K-1}]$—given small increments $\vec{X}$;
  $\overrightarrow{\eta^{(n)}} = \left[\eta_0^{(n)} \ldots \eta_{N-1}^{(n)}\right]$—is calculated by the formula (11) on n iterations of the method, based on the solution of the difference direct problem with respect to $\overrightarrow{X^{(n)}} = \left[X_0^{(n)}, \ldots, X_{K-1}^{(n)}\right]$ approximations;
- Moreover, $\overrightarrow{X^{(0)}}$ is the initial approximation for the components of $\vec{X}$;
- Similarly, we obtain $\overrightarrow{\eta^{(n)\delta}} = \left[\eta_0^{(n)\delta} \ldots \eta_{N-1}^{(n)\delta}\right]$ to solve the direct problem with respect to $\overrightarrow{X^{(n)\delta}} = \overrightarrow{X^{(n)}} + \overrightarrow{\delta X}$ approximations with given increments.

5 Algorithm of the Levenberg-Marquardt and Regularization

Remark 4 Typically, the so-called regularization parameter γ some number taking different values for different optimization problems, i.e., for different additional information $\vec{\theta}$.

The regularization parameter γ is of key importance for the Levenberg-Marquardt method and is directly related to the step and direction of convergence of the method. Therefore, it is worth noting important properties imposed on γ according to the:

$$\mathbb{Y} = \mathbb{R}_{>0} = \{\gamma | \gamma \in \mathbb{R}, \gamma > 0\};$$

- if $\gamma \in \mathbb{Y}$ and the Hesse matrix H is positively defined, then $\overrightarrow{\Delta X}$ is the direction of descent (gradient) for the "good" (optimal) step of the method;
- if $\gamma \to 0$ is very small, then $\overrightarrow{\Delta X}$ is the optimal step of the method in the gradient direction. This also means the convergence rate of the method is close to quadratic;
- If $\gamma \in \mathbb{Y}, \gamma \gg 0$, then it follows from (13) that $\overrightarrow{\Delta X} \simeq 1/ -{}^{1}\!/_{\gamma}\left(J^T \times \vec{\eta}\right)$ method step in the direction of the anti-gradient.

In this article we will not describe the Levenberg-Marquardt algorithm in detail; it can be found in detail in [42]. It is important to note only that there are two more parameters in the algorithm: c—is a given constant for recalculation of the regularization parameter and v—is a given constant, which is not necessary to compute the value of $\gamma^{(0)}$ at the start. The presented algorithm for solving the inverse problem (10) was implemented as a set of subroutines in C language.

Definition 5 As a result, the inverse problem for (4) is reduced to the choice of 4 parameters controlling the course of the iterative method, which are no longer tied to the physical meaning of the original task in specific models.

Remark 5 The peculiarity of the described methods of Newton-type minimization of the functional (12) for solving the difference inverse problem is that it is necessary to solve the difference direct problem (7) many times, different values of $\vec{X}$. This requires the most efficient implementations of the direct (7) problem and methods for solving it. We used the previously implemented subroutines [33] in C, using OpenMP [43] and/or CUDA [44], to speed up the calculations due to parallelization.

6 Test Examples and Results of Numerical Experiments

We will show by examples that the described method of solving the difference inverse problem (10), at certain given $\overrightarrow{X^{(0)}}, \delta X, c, v$ parameters related to the direction and step of the Levenberg-Marquardt method, really allows to restore the form of the functional dependence $\alpha(t)$. The structure of the numerical experiment is as follows:

Suppose for all examples further: $T = 30$, $N = 300$;

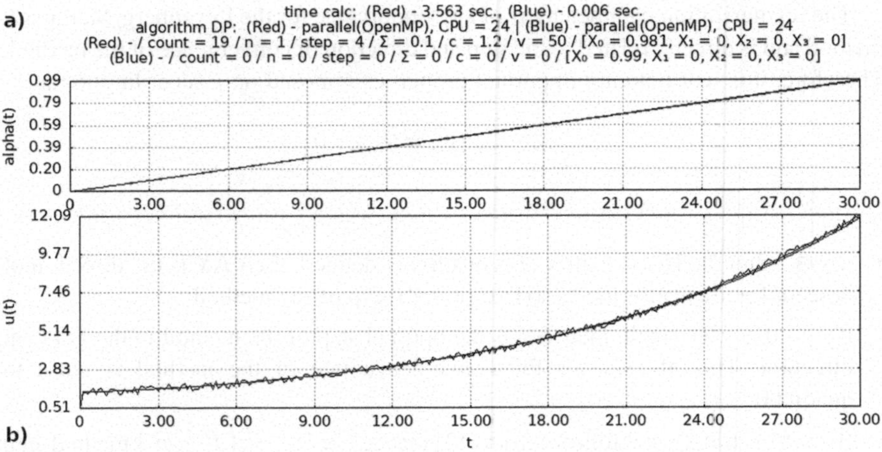

Fig. 1 Example 1. $\alpha(t_i) = X_0(i\tau/T)$—is a linear-increasing. (**a**) Graphs $\alpha(t_i)$: (Blue)—"benchmark" $\vec{X}$, (Red)—restored $\vec{X}$; (**b**) Model curves: (Blue)—"benchmark", (Red)—optimization result at $\overrightarrow{X^{(0)}} = [0.01]$, $\delta X = [0.005]$

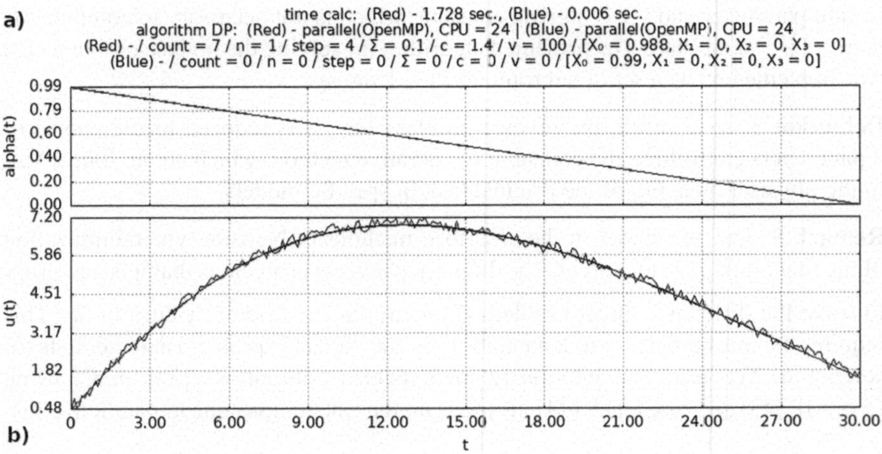

Fig. 2 Example 2. $\alpha(t_i) = X_0((N - i)\tau/T)$—linear-decreasing. (**a**) Graphs $\alpha(t_i)$: (Blue)—"benchmark" $\vec{X}$, (Red)—restored $\vec{X}$; (**b**) Model curves: (Blue)—"benchmark", (Red)—optimization result at $\overrightarrow{X^{(0)}} = [0.01]$, $\delta X = [0.005]$

- Let us solve the "benchmark" direct problem (7) with all given and known parameters of the model, including the recovering function $\alpha(t)$ and its defining parameters $\vec{X} = [X_0, \ldots, X_{K-1}]$. Blue model curve in (Figs. 1, 2, 3, 4, and 5b);
- Based on the "benchmark" solution, we generate $\vec{\theta}$—pseudo-random experimental data (8). The (Black) curve in (Figs. 1, 2, 3, 4, and 5b);

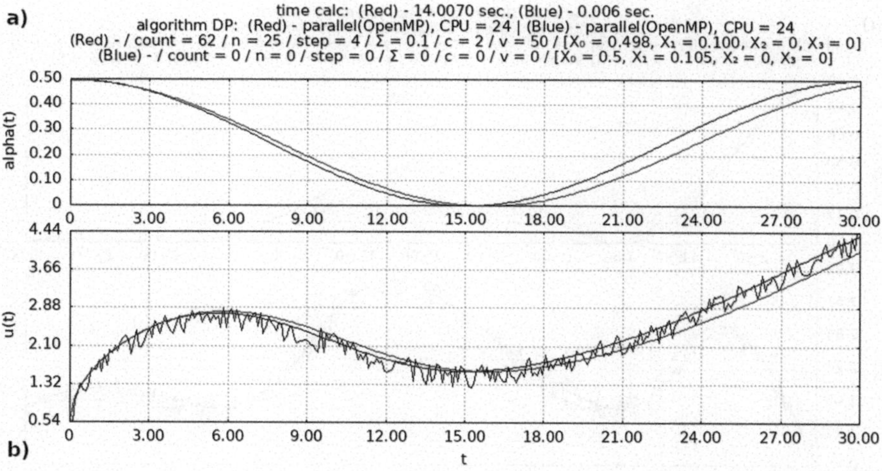

Fig. 3 Example 3. $\alpha(t_i) = X_0\cos^2(X_1 i\tau)$—periodic function. (**a**) Graphs $\alpha(t_i)$: (Blue)—"benchmark" $\vec{X}$, (Red)—restored $\vec{X}$; (**b**) Model curves: (Blue)—"benchmark", (Red)—optimization result at $\overrightarrow{X^{(0)}} = [0.1, 0.02]$, $\delta X = [0.05, 0.01]$

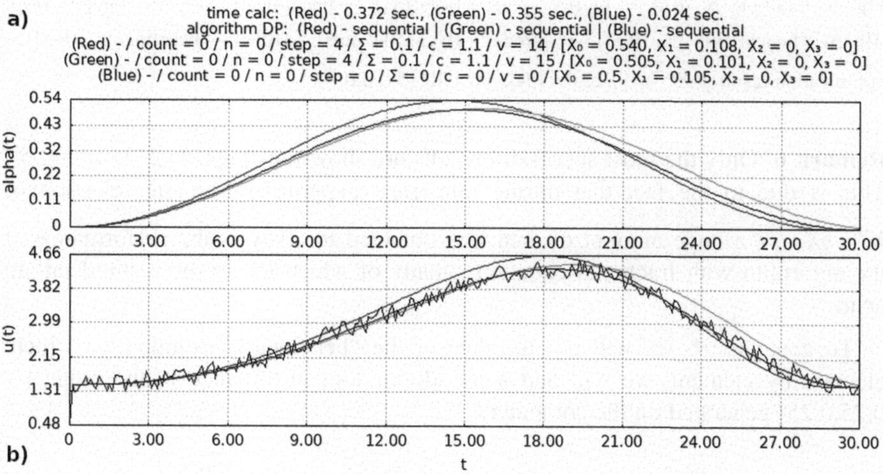

Fig. 4 Example 4. $\alpha(t_i) = X_0\sin^2(X_1 i\tau)$—periodic function. (**a**) Graphs $\alpha(t_i)$: (Blue)—"benchmark" $\vec{X}$, (Red)—restored $\vec{X}$; (**b**) Model curves: (Blue)—"benchmark", (Red)—optimization result at $\overrightarrow{X^{(0)}} = [0.005, 0.001]$, $\delta X = [0.005, 0.001]$

- Let us solve the difference inverse problem (10) by the described algorithm of the Levenberg-Marquardt method and restore the values of $\alpha(t) = \alpha\left(\vec{X}\right) = \alpha(X_0, \ldots, X_{K-1}) \in \widehat{A}$ according to $\vec{\theta}$. The (Red) and/or (Green) model curve in (Figs. 1, 2, 3, 4, and 5b).

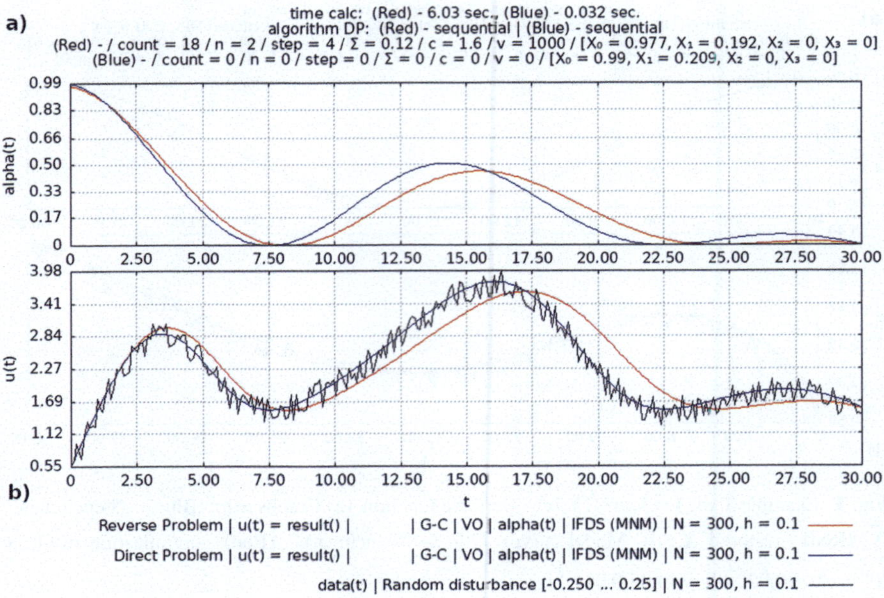

Fig. 5 Example 5. $\alpha(t_i) = (X_0((N - i)\tau/T))\cos^2(X_1 i\tau)$—linear-decreasing. (**a**) Graphs $\alpha(t_i)$: (Blue)—“benchmark” $\vec{X}$, (Red)—restored $\vec{X}$; (**b**) Model curves: (Blue)—“benchmark”, (Red)—optimization result at $\overrightarrow{X^{(0)}} = [0.22, 0.0404]$, $\delta X = [0.15, 0.03]$

Remark 6 Only the most successful results are shown in (Figs. 1, 2, 3, 4, and 5). This is due to the fact that during numerical experiments on the selection of $\overrightarrow{X^{(0)}}, \delta X, c, v$ a huge amount of data was obtained to analysis the performance of the algorithm with fractional equations, many of which led to the method falling apart.

To generate $\vec{\theta}$ we will use the data of the “benchmark” solution, to which, element by element, we will add a pseudo-random disturbance in the range [−0.25,0.25] generated on the computer:

$$\theta(t_i) = u(t_i) + p(s), \quad i = 0..N-1, \quad s = \pm 0.25,$$

where, $p(s)$—is the distribution function of the discrete random variable s. This method of generating experimental data is implemented in the C programming language on the basis of rand()—function for generating pseudo-random numbers, standard library <stdlib.h>.

Calculations for solution direct and inverse problems in test examples were carried out on a personal computer purchased for work under the grant of the Russian Science Foundation № 23-71-01050 on the theme “Development of a software package for modeling and analysis of volumetric activity of radon as a precursor of strong earthquakes in Kamchatka”. The personal computer has the following system characteristics: CPU—AMD Ryzen 97950X, 16 × 4.5 GHz (32 Threads); RAM—96 Gb; GPU—GeForce RTX 4090, 24 Gb, 2235 MHz, ALU 16384.

7 Conclusion

From the results we can conclude that it is possible to solve the inverse problem for an in-homogeneous fractional equation with a differentiation operator of Gerasimov-Caputo type of variable order. It is shown that with the help of mathematical unconditional optimization methods, in particular the iterative Levenberg-Marquardt method, having a mathematical model and experimental data of the process, it is possible to recover close to optimal values of $\alpha(t)$ the fractional derivative degree, including as a function of $\alpha(t)$ from two variables.

However, the more complex the structure of the experimental data and the type of function defining $\alpha(t)$ are, the more difficult it is to choose the parameters controlling the solution of the inverse problem of the Levenberg-Marquardt method and the starting values of $\overrightarrow{X^{(0)}}, \delta X, c, v$ at which the conditions imposed on the regularization parameter γ would be satisfied and method would converge to the optimal solution.

The continuation of this work consists in passing from the solution of inverse problems for the in-homogeneous fractional equation to inverse problems for the quadratic nonlinear fractional analogue of the Riccati equation. In turn, this will allow us to solve various inverse problems for the determination of certain parameters of the model equations of the volumetric activity of radon gas on the basis of recorded experimental data, which is of practical interest.

Funding The research was funded by a grant from the Russian Science Foundation, project number 23-71-01050, which can be found at https://rscf.ru/project/23-71-01050/.

References

1. Kilbas, A.A., Srivastava, H.M., Trujillo, J.J.: Theory and Applications of Fractional Differential Equations. Elsevier, Amsterdam (2006)
2. Iomin, A.: Fractional-time quantum dynamics. Phys. Rev. E. **80**(2), 1–4 (2009)
3. Bagley, R.L., Torvik, P.J.: A theoretical basis for the application of fractional calculus to viscoelasticity. J. Rheol. **27**(3), 201–210 (1983)
4. Coimbra, C.F.M.: Mechanics with variable-order differential operators. Ann. Phys. **12**(11–12), 692–703 (2003)
5. Rossikhin, Y.A., Shitikova, M.V.: Application of fractional calculus for dynamic problems of solid mechanics: novel trends and recent results. Appl. Mech. Rev. **63**(1:010801), 1–52 (2010)
6. Moroz, L.I., Maslovskaya, A.G.: Numerical simulation of an anomalous diffusion process based on a scheme of a higher order of accuracy. Math. Models Comp. Simulat. **13**(3), 492–501 (2021)

7. Sun, H.G., Chen, W., Wei, H., Chen, Y.Q.: A comparative study of constant-order and variable-order fractional models in characterizing memory property of systems. Eur. Phys. J. Special Topics. **193**(1), 185–192 (2011)
8. Volterra, V.: Sur lese quations integro-differentielles et leurs applications. Acta Mathematica. **35**(1), 295–356 (1912)
9. Nakhushev, A.M.: Fractional Calculus and Its Application. Fizmatlit, Moscow (2003)
10. Rekhviashvili, S.S., Pskhu, A.V.: Fractional oscillator with exponential-power memory function. Tech. Phys. Lett. **48**(7), 33–35 (2022)
11. Gerasimov, A.N.: Generalization of linear deformation laws and their application to internal friction problems. Appl. Math. Mech. **12**, 529–539 (1948)
12. Caputo, M.: Linear models of dissipation whose Q is almost frequency independent—II. Geophys. J. Int. **13**(5), 529–539 (1967)
13. Uchaikin, V.V.: Fractional Derivatives for Physicists and Engineers Background and Theory, vol. I. Springer, Berlin (2013)
14. Patnaik, S., Hollkamp, J.P., Semperlotti, F.: Applications of variable-order fractional operators: a review. Proc. R. Soc. A. **476**(2234), 20190498 (2020)
15. Fang, Z.W., Sun, H.W., Wang, H.: A fast method for variable-order Caputo fractional derivative with applications to time-fractional diffusion equations. Comput. Math. Appl. **80**(5), 1443–1458 (2020)
16. Sahoo, S., Saha Ray, S., Das, S., Bera, R.K.: The formation of dynamic variable-order fractional differential equation. Int. J. Modern Phys. C. **27**(07), 1650074 (2016)
17. Tverdyi, D.A., Parovik, R.I.: Investigation of finite-difference schemes for the numerical solution of a fractional nonlinear equation. Fractal Fract. **6**(1:23), 1–27 (2022)
18. Tverdyi, D.A., Makarov, E.O., Parovik, R.I.: Hereditary Mathematical Model of the Dynamics of Radon Accumulation in the Accumulation Chamber. Mathematics. **11**(4:850), 1–20 (2023)
19. Tverdyi, D.A., Makarov, E.O., Parovik, R.I.: Research of stress-strain state of geo-environment by emanation methods on the example of alpha(t)-model of radon transport. Bulletin KRASEC. Phys. Math. Sci. **44**(3), 86–104 (2023)
20. Zuzel, G., Simgen, H.: High sensitivity radon emanation measurements. Appl. Radiat. Isot. **67**(5), 889–893 (2009)
21. Cox, D.R., Hinkley, D.V.: Theoretical Statistics. Chapman & Hall/CRC, London (1979)
22. Reviznikov, D.L., Morozov, A.Y.: Algorithms for the numerical solution of fractional differential equations with interval parameters. Siberian J. Ind. Math. **26**(4), 93–108 (2023)
23. Slastushenskiy, Y.V., Reviznikov, D.L., Semenov, S.A.: Methods for parametric identification of fractional differential equations. Diff. Eq. **60**(7), 954–966 (2024)
24. Hadamard, J.S.: Sur les problèmes aux derivees partielles et leur significa tion physique. Princeton Univ. Bull. **13**(4), 49–52 (1902)
25. Morozov, V.A.: Methods for Solving Incorrectly Posed Problems. Springer, New York, NY (1984)
26. Mueller, J.L., Siltanen, S.: Linear and Nonlinear Inverse Problems with Practical Applications. Society for Industrial and Applied Mathematics, Philadelphia, PA (2012)
27. Tarantola, A.: Inverse Problem Theory: Methods for Data Fitting and Model Parameter Estimation. Elsevier Science Pub. Co., Amsterdam (1987)
28. Tvoyrdyj, D.A.: The Cauchy problem for the Riccati equation with variable power memory and non-constant coefficients. Bulletin KRASEC. Phys. Math. Sci. **23**(3), 148–157 (2018)
29. Tahmasebi, P., Javadpour, F., Sahimi, M.: Stochastic shale permeability matching: three-dimensional characterization and modeling. Int. J. Coal Geol. **165**(1), 231–242 (2016)
30. Lailly, P.: The seismic inverse problem as a sequence of before stack migrations. Conference on Inverse Scattering, Theory and application, 206–220 (1983)
31. Mohamad-Djafari, A.: Inverse Problems in Vision and 3D Tomography. ISTE-WILEY, New-York, NY (2010)
32. Hayotov, A.R., Jeon, S., Shadimetov, K.M.: Application of optimal quadrature formulas for reconstruction of CT images. J. Comput. Appl. Math. **388**, 113313 (2021)

33. Tverdyi, D.A., Parovik, R.I.: Application of high-performance computing to solve the Cauchy problem with the fractional Riccati equation using an nonlocal implicit finite-difference scheme. Bulletin KRASEC. Phys. Math. Sci. **46**(1), 103–117 (2024)
34. Tikhonov, A.N., Samarskii, A.A.: Equations of Mathematical Physics. Macmillan, London (1963)
35. Dennis, J.E., Robert, J., Schnabel, B.: Numerical Methods for Unconstrained Optimization and Nonlinear Equations. SIAM, Philadelphia, PA (1983)
36. Tikhonov, A.N.: On the solution of ill-posed problems and the method of regularization. Dokl. Akad. Nauk. **151**(3), 501–504 (1963)
37. Kabanikhin, S.I., Iskakov, K.T.: Optimization Methods for Solving Coefficient Inverse Problems. NGU, Novosibirsk (2001) (In Russian)
38. Dahlquist, G., Bjorck, A.: Numerical Methods. Prentice-Hall, Hoboken, NJ (1974)
39. Arridge, S.R., Schweiger, M.: A general framework for iterative reconstruction algorithms in optical tomography, using a finite element method. Comput. Radiol. Imaging Therapy Diagn. **110**, 40–70 (1999)
40. Levenberg, K.: A method for the solution of certain non-linear problems in least squares. Q. Appl. Math. **2**(2), 164–168 (1944)
41. Marquardt, D.W.: An algorithm for least-squares estimation of nonlinear parameters. J. Soc. Ind. Appl. Math. **11**(2), 431–441 (1963)
42. More, J.J.: The Levenberg-Marquardt algorithm: implementation and theory. In: Watson, G.A. (ed.) Numerical Analysis Lecture Notes in Mathematics, vol. 630, pp. 105–116. Springer, Berlin (1978)
43. Supinski, B., Klemm, M.: OpenMP Application Programming Interface Specification Version 5.2. Independently Published, North Charleston (2021)
44. Sanders, J., Kandrot, E.: CUDA by Example: an Introduction to General-Purpose GPU Programming. Addison-Wesley Professional, London (2010)

Computational Approach to Modeling Nutrient-Dependent Growth of Bacterial Patterns with Time Delay Effect

Ivan Shevkun, Anna Maslovskaya, and Christina Kuttler

Abstract The paper is devoted to developing a mathematical model of nutrient-dependent bacterial growth, modified by incorporating a time delay effect. The model is formalized by an initial-boundary value problem for a system of reaction-diffusion nonlinear partial differential equations, taking into account the time lag between nutrient consumption by bacteria and their subsequent reproduction. A computational scheme based on the implicit finite difference method was constructed. The numerical algorithm was supplemented with a stochastic procedure of deformation providing naturalistic growth patterns of bacterial colonies. A series of computational experiments are conducted to explore the spatio-temporal distributions of bacterial biomass concentration depending on the lag effect. Computational experiments allow us to reveal the specification of the dependence of the bacterial growth rate on the time delay control parameter.

Keywords Bacterial growth · Bacterial patterns · Reaction-diffusion system · Time delay model · Finite difference method · Simulation of dendrite structures

1 Introduction

At present, computational and mathematical biophysics are promising and intensively developing areas of research that are driving progress in elucidating mechanisms and solving open problems at the forefront of the life sciences, particularly in microbiology. A particular focus of such research is *in silico* studies, which allow a

I. Shevkun
Innopolis University, Innopolis, Russia

A. Maslovskaya
Innopolis University, Innopolis, Russia

C. Kuttler (✉)
Technical University of Munich, Garching, Germany
e-mail: kuttler@ma.tum.de

A. Gibadullin, S. Gordin (eds.), *Computing Technologies and Applied Mathematics*, Springer Proceedings in Mathematics & Statistics 500,
https://doi.org/10.1007/978-3-031-90546-9_12

biological object to be studied using computer simulation software. In particular, *in silico* models combine differential equations with powerful computing resources to perform computational experiments in addition to or instead of in vivo experiments. Understanding the mechanisms that regulate the behavior of bacterial communities, the most important representatives of the micro-world, is an extremely challenging problem for both, basic and applied research.

The classic method of microbiological diagnosis and detection of infectious diseases is the cultivation of microorganisms using special nutrient substrates. Under certain cultivation conditions, bacterial colonies are evolving self-similar structures with a dendrite-like structure [1]. Currently, several basic deterministic approaches to modeling the spatio-temporal dynamics of bacterial biomass are known [2–4]. In addition, discrete dynamical simulations have also been the focus of intense research efforts. These classes of models are based on the application of various approaches using stochastic Monte Carlo simulations, agent-based modeling using cellular automaton methodology, and the particle method (see for example [5, 6] and references therein). The development of models of bacterial evolution is of particular importance due to the ability of many virulent bacterial species to act collectively through intercellular communication and form factors of tolerance to the action of antibacterial medicines [7]. Biomass evolution models form an important part of bacterial quorum models, which have been intensively developed recently [8–11].

In the context of the present study, the emphasis is on using a deterministic approach and corresponding application of differential approaches to model spatio-temporal characteristics of biomass as well as nutrient concentrations during the evolution of bacterial colonies. Differential equation models represent one of the most studied and tested classes of approaches used for formalizing bacterial growth in nutrient media and formation of dendrite patterns. Fundamental well-known model approaches using the apparatus of ordinary differential equations are the Monod and Droop terms [12, 13]. The introduction of spatio-temporal characteristics of bacterial biomass and nutrient concentration based on the formalization of partial differential equations represents a natural continuation and expansion of the deterministic setting. Here we especially highlight the class of "reaction-diffusion" type models: the Kolmogorov-Petrovsky-Piskunov-Fisher model [14], the Kessler and Levin model [15], the Kawasaki model [2], the spatio-temporal modification of the Monod model [4], fractional differential models [16, 17], and various model modifications, including chemotaxis models [18–20].

It should be noted that delayed feedback is one of the most important regulatory systems for biological processes [21]. Many biological objects do not respond immediately to external changes, but with a certain time lag. There is previous experimental evidence that when the concentration of nutrients changes, the rate of consumption by bacteria is immediate, but changes in the growth rate of the bacterial mass are delayed [22]. To be precise, this time lag is estimated to be about 20 min for Escherichia coli [22]. In principle, the time difference between cause and effect in a dynamical system can be formalized by introducing a time lag into the differential equations governing the model. Accordingly, for models described by ordinary differential equations (ODE), the mathematical formulation will lead to delayed differential equations (DDE), while for models formalized by reaction-diffusion partial differential equations (PDE), the diffusion term, the reaction part, the source component, or the boundary conditions can take into account the time delay.

In the framework of the present study following [22], we assume that there is a time delay between nutrient consumption by bacterial colonies and bacterial growth. In this way, the purpose of the study was to develop a mathematical model of nutrient-dependent bacterial growth, modified by the inclusion of a time delay effect, focusing on the application of combined numerical methods and computer simulation techniques.

2 Materials and Methods

In order to construct the time delay modification of a model of bacterial growth and derive a numerical scheme, let us give a short explanation of considered biological processes according to the system approach.

2.1 *Brief Biological Setup*

For instance, we will consider a model representation of the surface growth of bacterial colonies cultured in nutrient media (nutrient agar is the most commonly used nutrient media for the cultivation of microbes in petri dishes and detection of pathogens). We also assume that, as a result of seeding bacteria at a given point on a two-dimensional square surface (simulating, for example, a Petri dish), a colony of bacteria begins to develop, continuously consuming nutrients. Figure 1 shows the functional diagram of the biosystem. We suppose that the spatio-temporal distributions of biomass concentration b and nutrient concentration n should be specified.

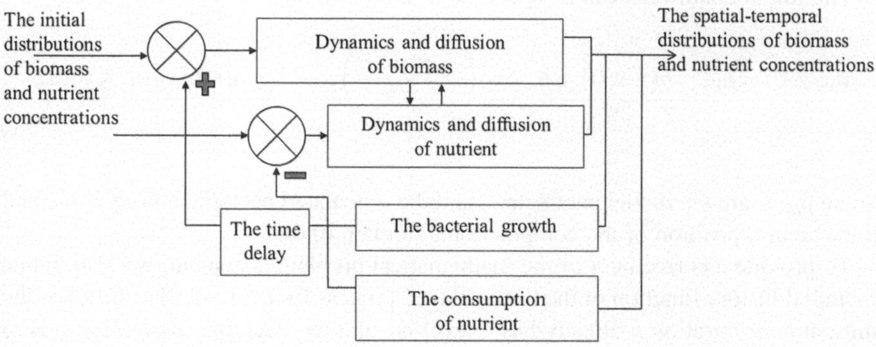

Fig. 1 Model structure for bacterial growth on a nutrient medium with a time delay effect

According to above, we will also assume that the biomass concentration changes with a certain time delay α.

The systems approach allows us to arrive at a mathematical model of nutrient-dependent growth of bacterial patterns, which can be expressed by a system of time-dependent reaction-diffusion equations.

2.2 Mathematical Formulation of the Problem

A mathematical model of bacterial growth allows us to quantify the spatio-temporal dynamics of biomass and nutrient substrate concentrations. In this study, we introduce a modification of the model proposed in [2] by incorporating a time delay effect. The basic model assumes nonlinear dependence of biomass diffusion on the nutrient concentration. The generation term corresponds to the modified representation of Monod kinetics. Here, considering the 2D model and taking into account the normalization of the variables, we formulate a problem of evaluation of the spatio-temporal distributions of biomass $b(x,y,t)$ and nutrient concentration $n(x,y,t)$ in square computational domain with the linear size of 2 L. The normalized model modified by incorporating a time delay effect between the consumption of nutrients by bacteria and their growth, is formalized as an initial boundary value problem for a nonlinear system of partial differential equations with delay:

$$\begin{cases} \dfrac{\partial n}{\partial t} = \Delta n - v_n bn, \\ \dfrac{\partial b}{\partial t} = \nabla(\sigma nb\nabla b) + v_b n(t-\alpha)b(t-\alpha), \end{cases} \tag{1}$$

where $-L < x < L$, $-L < y < L$, $0 < t \leq T$; $\sigma = \sigma_0\ (1 + \delta)$ is the model parameter characterizing the nutrient concentration; v_n, v_b are positive parameters; δ is a random variable responsible for fluctuations in bacterial mass during colony growth; $|\delta| < 1$, where α is the time delay, $\alpha \geq 0$.

The initial conditions can be specified in the following forms:

$$n(x,y,0) = n_0, \quad b(x,y,0) = \beta_m \exp\left(-\frac{x^2+y^2}{C^2}\right), \quad -L \leq x \leq L, \quad -L \leq y \leq L, \tag{2}$$

where β_m, C are the model parameters, and the inoculated bacterial colony is located at the central position of the computational domain (0,0).

To provide a correctness of the mathematical problem statement, we also define the initial history function of the time-delayed process for $t \in [-\alpha, 0)$ as follows: the nutrient concentration n already has a level n_0, and the bacterial mass b has a zero value throughout the entire computational domain.

Furthermore, we impose the natural boundary conditions

$$\left.\frac{\partial n}{\partial \mathbf{n}}\right|_{\Gamma} = 0, \quad \left.\frac{\partial b}{\partial \mathbf{n}}\right|_{\Gamma} = 0, \quad 0 < t \leq T, \tag{3}$$

where Γ is the boundary of the computational domain with linear dimensions $2\,L$, $\mathbf{n}$ is the external normal vector to the boundary Γ.

Note that the question of theoretical research of basic models for global and local solvability is a separate scientific problem. In a generalized formulation (with the representation of the nonlinear diffusion coefficient in the form $\sigma n^{\gamma} b^{(\beta + 1)}$, where α and β are positive integer constants and for a fixed constant value σ), this problem has been analyzed in [23]. The construction of the solution of reaction-diffusion problems in non-linear formulations poses particular difficulties. Therefore, numerical methods play an important role in the practice of applied modelling. Within the framework of this study, we applied the implicit finite-difference splitting method.

It should be noted that the issue of theoretical justification of the correctness of mathematical formulations of delay problems is rather complicated and requires separate consideration. In the practice of computer modelling of delayed systems, researchers often assume that the solution to the problem exists and is unique, then derive numerical schemes and either prove the convergence of such schemes on the basis of a rigorous analysis, or perform numerical testing and verification of practical convergence on the basis of Runge's rule [24]. Among the set of numerical methods for solving spatial evolutionary problems with delay, the group of finite difference methods is the most developed and theoretically studied. The finite difference method is simple in the construction of the computational scheme and its implementation, but at the same time it allows to achieve the required accuracy by reducing the grid step in coordinates and/or time. In our problem, the coordinate solution domain is a simple rectangular domain, which also favors the finite difference method. Finally, the regular grid of the finite difference method allows easy visualization of the solution, which is also important when considering the geometric patterns of bacterial colonies.

2.3 Numerical Scheme

According to the fractional step scheme, the solution process is divided into two half-steps in time, each of which considers the derivative in only one spatial direction, x or y. In our case, we present a computational scheme for the second equation of the system (3). The second equation, which defines the space-time distribution of biomass density is written in a more complex form due to nonlinear dependence of the diffusion coefficient as well as a time delay for the reaction term. At the first temporal half-step, the finite difference scheme is given as follows:

$$\frac{b^{k+1/2}-b^{k}}{\tau}=\sigma n_{i,j}^{k+1/2}b_{i,j}^{k}\frac{b_{i+1,j}^{k+1/2}-2b_{i,j}^{k+1/2}+b_{i-1,j}^{k+1/2}}{h_x^2}+$$
$$+\sigma\left(\frac{n_{i+1,j}^{k+1/2}-n_{i-1,j}^{k+1/2}}{2h_x}b_{i,j}^{k}+n_{i,j}^{k+1/2}\frac{b_{i+1,j}^{k}-b_{i-1,j}^{k}}{2h_x}\right)\frac{b_{i+1,j}^{k+1/2}-b_{i-1,j}^{k+1/2}}{2h_x}$$
$$+\frac{1}{2}n_{i,j}^{k-q+1/2}b_{i,j}^{k-q}, \tag{4}$$

where $i=\overline{1,N-1}$, $j=\overline{1,M-1}$, $k=\overline{0,P-1}$, and the index q corresponds to the delay time $\alpha_q = q\tau$.

For the second temporal half-step we arrive at the following approximation:

$$\frac{b^{k+1}-b^{k+1/2}}{\tau}=\sigma n_{i,j}^{k+1}b_{i,j}^{k+1/2}\frac{b_{i,j+1}^{k+1}-2b_{i,j}^{k+1}+b_{i,j-1}^{k+1}}{h_y^2}+$$
$$+\sigma\left(\frac{n_{i,j+1}^{k+1}-n_{i,j-1}^{k+1}}{2h_y}b_{i,j}^{k+1/2}+n_{i,j}^{k+1}\frac{b_{i,j+1}^{k+1/2}-b_{i,j-1}^{k+1/2}}{2h_y}\right)\frac{b_{i,j+1}^{k+1}-b_{i,j-1}^{k+1}}{2h_y}$$
$$+\frac{1}{2}n_{i,j}^{k-q+1}b_{i,j}^{k-q+1/2}, \tag{5}$$

where $i=\overline{1,N-1}$, $j=\overline{1,M-1}$, $k=\overline{0,P-1}$, $n_{i,j}^{k}$ and $b_{i,j}^{k}$ are the values of the functions n and b at the approximation node (i,j,k), respectively.

The systems of linear equations obtained as a result of the finite difference approximation were solved by the double-sweep method. The numerical scheme was complemented by a stochastic procedure for calculating the deformation parameter σ for each time step, which allows the simulation of naturalistic growth patterns of bacterial colonies. To determine the total volume of bacterial mass B at a specific time step k, we summed the b values over all approximation nodes:

$$B^k=\sum_{i=0}^{N}\sum_{j=0}^{M}b_{i,j}^{k}. \tag{6}$$

The computational scheme was implemented as an application program using Matlab software.

3 Results and Discussion

We present the results of mathematical modelling within the framework of several computational experiments. The main purpose is to visualize patterns of bacterial cultures of a branching morphological type and to study the dependence of growth phenomena on the delay time. In a developed software application, the computational experiments involve the initialization of a set of normalized model parameters and the setting of computational parameters. For the calculations, the following values of the parameters were set (in arbitrary units) $\nu_n = 1$, $\nu_b = 1$, $\beta_m = 0.71$, $C^2 = 6.25$, $n_0 = 0.5$, $\sigma_0 = 1$. The value of the random parameter δ was generated at each approximation node using a triangular distribution in the range $(-1, 1)$ to give a dendritic branching shape to the solution. The observation time T was chosen empirically for each parameter set (so that the total value of the bacterial mass reaches a certain value B_m). The results of the program are spatio-temporal distributions of nutrient and biomass concentrations. Figure 2 shows space distributions of bacterial biomass at fixed time moments for different time delays (computations stopped at defined value of B_m).

The obtained findings indicate that taking into account even a small time delay, we result in significant changes in the simulation results. The growth of a bacterial colony on a nutrient medium slows down significantly due to the time delay effect. If we compare the simulation data with the data from a biological experiment for the bacterial species *Pseudomonas putida* [25], we can approximate the correspondence between the bacterial growth time in the computational experiment and the colony growth time, and as a consequence, for a particular experiment we can match the conditional normalized time delay and the real delay time. Specifically, if a colony reaches a linear size of 0.5 cm in 8 days (let us choose Fig. 2b for example), then the normalized time $t^* = 1$ a.u. is related to 0.133 h, then the time delay of $\alpha = 5$ a.u. corresponds to 40 min.

To analyze the influence of the delay effect on the character of bacterial growth in more detail, we calculated the total amount of biomass B defined by relation (6), during the dynamics of the bacterial growth, as shown in Fig. 3 (note that the simulation time interval is limited to the period of intensive bacterial growth and does not include the stationary phase). These data suggest the specification of a time delay effect during nutrient digestion, which leads to a slowdown in the growth of the bacterial colony. Specifically, this effect is more pronounced at low values of time delay, but as the time delay increases, the growth slowdown is not as pronounced as at the beginning. For example, the decrease in growth rate between time delay values of 0 and 5 a.u. is approximately the same as between values of 5 a.u. and 40 a.u.

4 Conclusion

Thus, the study generalizes the result of the development of an approach to mathematically model the growth of bacterial colonies under surface culture on nutrient media. The new approach presented a differential equation model of nutrient-dependent bacterial growth, modified by the inclusion of a time delay effect. To solve the problem numerically, we proposed a computational scheme based on a combination of implicit finite difference methods and a procedure for calculating the

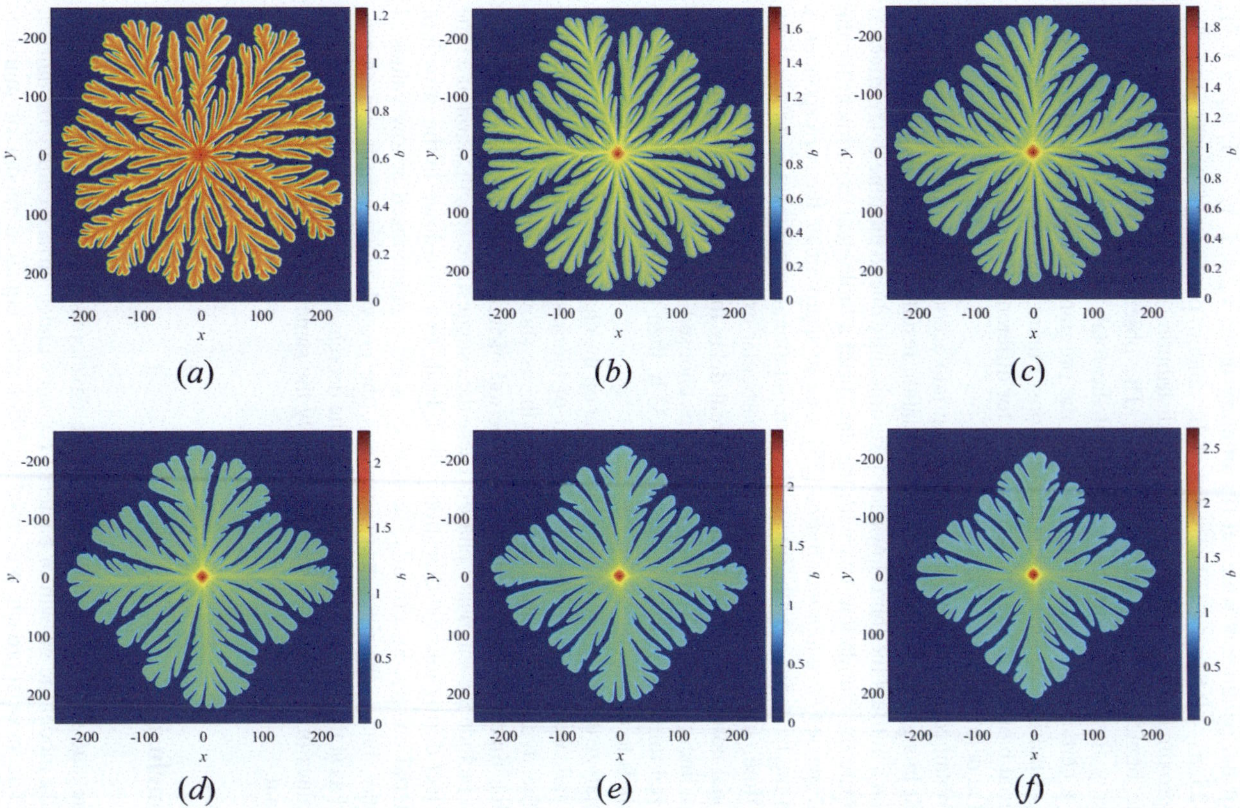

Fig. 2 The simulation results of bacterial patterns under variation of time delay fixed at different time moments: (**a**)—$\alpha = 0$ for $t = 3400$, (**b**)—$\alpha = 5$ for $t = 5500$, (**c**)—$\alpha = 10$ for $t = 7500$, (**d**)—$\alpha = 20$ for $t = 10{,}000$, (**e**)—$\alpha = 30$ for $t = 12{,}000$, (**f**)—$\alpha = 40$ for $t = 12{,}600$

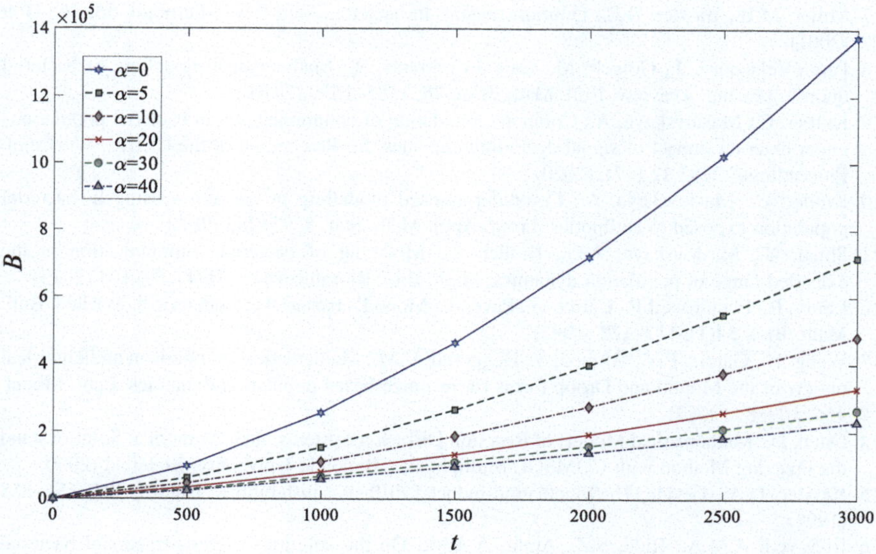

Fig. 3 Temporal dependence of bacterial mass B calculated for different values of time delay

stochastic diffusion parameter. This approach allows to visualize naturalistic bacterial patterns that characterize the geometrical structure of the dendrites as well as to estimate the influence of the time delay on the bacterial growth rate. Additional studies of key characteristics of bacterial growth under variations of inhibitory regimes of antibiotic action require further investigation.

Acknowledgments The study has been supported by the Ministry of Science and Higher Education of Russia (Project no. 122082400001-8).

References

1. Xiong, L., Cao, Y., Cooper, R., Rappel, W.-J., Hasty, J., Tsimring, L.: Flower-like patterns in multi-species bacterial colonies. eLife. **9**, e48885 (2020)
2. Kawasaki, K., Mochizuki, A., Matsushita, M., Umeda, T., Shigesada, N.: Modeling spatio-temporal patterns generated by *Bacillus subtilis*. J. Theor. Biol. **188**(2), 177–185 (1997)
3. Mimura, M., Sakaguchi, H., Matsushita, M.: Reaction-diffusion modelling of bacterial colony patterns. Physica A. **282**(1–2), 283–303 (2000)
4. Eberl, H.J., Parker, D.F., van Loosdrecht, M.C.M.: A new deterministic spatio-temporal continuum model for biofilm development. Comput. Math. Meth. Med. **3**, 161–175 (2001)
5. Delavar, M.A., Wang, J.: Advanced Methods and Mathematical Modeling of Biofilms. Applications in Health Care, Medicine, Food, Aquaculture, Environment, and Industry. Academic Press, Elsevier (2022)
6. Vassallo, L., Hansmann, D., Braunstein, L.A.: On the growth of non-motile bacteria colonies: an agent-based model for pattern formation. Eur. Phys. J. B. **92**, 216 (2019)

7. Miller, M.B., Bassler, B.L.: Quorum sensing in bacteria. Ann. Rev. Microbiol. **55**, 165–199 (2001)
8. Perez-Velazquez, J., Gölgeli, M., Garcia-Contreras, R.: Mathematical modelling of bacterial quorum sensing: a review. Bull. Math. Biol. **78**, 1585–1639 (2016)
9. Kuttler, C., Maslovskaya, A.: Computer simulation of communication in bacterial populations under external impact of signal-degrading enzymes. In: Proceeding of the CEUR "Workshop Proceedings" 2783 12 (17), (2020)
10. Kuttler, C., Maslovskaya, A.: Computer-assisted modelling of quorum sensing in bacterial population exposed to antibiotics. Front. Appl. Math. Stat. **8**, 951783 (2022)
11. Shuai, Y., Maslovskaya, A.G., Kuttler, C.: Modeling of bacterial communication in the extended range of population dynamics. Math. Biol. Bioinformatics. **18**(1), 89–104 (2023)
12. Lobry, R., Flandrois, J.P., Carret, G., Pave, A.: Monod's bacterial growth model revisited. Bull. Math. Biol. **54**(1), 117–122 (1992)
13. Wang, H., Garcia, P.V., Ahmed, S., Heggerud, C.M.: Mathematical comparison and empirical review of the Monod and Droop forms for resource-based population dynamics. Ecol. Model. **466**, 109887 (2022)
14. Otten, D.: Mathematical Models of Reaction Diffusion Systems, their Numerical Solutions and the Freezing Method with COMSOL Multiphysics. Bielefeld University, Bielefeld (2000)
15. Kessler, D.A., Levine, H.: Fluctuation-induced diffusive instabilities. Nature. **394**, 556–558 (1998)
16. El-Sayed, A.M.A., Rida, S.Z., Arafa, A.A.M.: On the solutions of time-fractional bacterial chemo-taxis in a diffusion gradient chamber. Int. J. Nonlinear Sci. **7**, 485–492 (2009)
17. Moroz, L.I.: Time-fractional numerical modelling applied to diffusion-wave processes of bacterial biomass growth. Far Eastern Math. J. **22**(2), 207–212 (2022)
18. Rida, S.Z., El-Sayed, A.M.A., Arafa, A.A.M.: Effect of bacterial memory dependent growth by using fractional derivatives reaction-diffusion chemotactic model. J. Stat. Phys. **140**, 797–811 (2010)
19. Kitsunezaki, S.: Interface dynamics for bacterial colony formation. J. Phys. Soc. Jpn. **66**, 1544–1550 (1997)
20. Golding, I., Kozlovsky, Y., Cohen, I., Ben-Jacob, E.: Studies of bacterial branching growth using reaction-diffusion models for colonial development. Physica A. **260**, 510–554 (1998)
21. Milton, J.G.: Time delays and the control of biological systems: an overview. IFAC-PapersOnLine. **48**(12), 87–92 (2015)
22. Ellermeyer, S., Hendrix, J., Ghoochan, N.A.: Theoretical and empirical investigation of delayed growth response in the continuous culture of bacteria. J. Theor. Biol. **222**(4), 485–494 (2003)
23. Horger, T., Kuttler, C., Wohlmuth, B., Zhigun, A.: Analysis of a bacterial model with nutrient-dependent degenerate diffusion. Math. Methods Appl. Sci. **38**(17), 3851–3865 (2015)
24. Lekomtsev, A.V., Pimenov, V.G.: Convergence of the alternating direction method for the numerical solution of a heat conduction equation with delay. Proc. Steklov Inst. Math. **272**(Suppl. 1, 04), S101–S118 (2011)
25. Shapiro, J.A.: Control of *Pseudomonas putida* growth on agar surfaces. In: Sokatch, J.R. (ed.) The Bacteria X. Academic Press, New York (1986)

A Combined Approach to the Classification and Semantic Segmentation of Production Tools Through the Use of the One-Shot Learning Method

D. M. Grabar, S. V. Zhiganov, and Y. S. Ivanov

Abstract The paper addresses the issue of tool segmentation and classification in an automated manufacturing environment by employing the One-Shot Learning technique. A novel architecture is proposed that integrates Siamese neural networks with a pretrained ConvNeXt model for feature extraction, enabling efficient classification tasks with minimal data input. The distinctive aspect of this approach is the employment of the LRASPP semantic segmentation technique, which enhances the precision of object identification in images. A comparative analysis of traditional and contemporary One-Shot Learning approaches is conducted, and the efficacy of the suggested architecture is validated through testing on a custom dataset comprising four distinct tool classes. Through experimentation, a classification accuracy rate of 96.87% is achieved, demonstrating the high efficacy and dependability of the proposed methodology. The findings demonstrate that the proposed system has the potential to enhance the efficiency and precision of operator monitoring in the production process, thereby reducing the likelihood of errors and defects.

Keywords One-shot learning · Segmentation · Classification · Automated systems · Dataset

1 Introduction

Modern industrial production is characterized by a high level of automation, extensive human interaction with automated systems, and the development of synergy between human operators and robots, thanks to advances in collaborative robotics and artificial intelligence. Under these conditions, it is essential to ensure accurate interpretation of operator actions through video. This not only enhances operational efficiency but also facilitates timely detection of errors that could lead to manufacturing defects.

D. M. Grabar (✉) · S. V. Zhiganov · Y. S. Ivanov
Komsomolsk-on-Amur State University, Komsomolsk-on-Amur, Russian Federation
e-mail: grabardm@ml-dev.ru

A. Gibadullin, S. Gordin (eds.), *Computing Technologies and Applied Mathematics*, Springer Proceedings in Mathematics & Statistics 500,
https://doi.org/10.1007/978-3-031-90546-9_13

Modern methods of video data analysis enable a deeper understanding and interpretation of human behavior in the production setting, presenting new opportunities to optimize technological processes and improve efficiency.

The use of the techniques for semantic annotation of videos presented in [1, 2] allows for more accurate identification and classification of the actions of the operator. This is especially significant in dynamic production environments, where the human factor can be a significant contributor to errors.

When human error is a key cause of defects, it is essential not only to record the actions of personnel, but also to comprehend their context. Research [3, 4] demonstrates the efficacy of advanced video analysis techniques for recognizing and interpreting intricate occurrences. These approaches permit not only the identification of errors, but also the formulation of strategies to prevent them.

The integration of advanced video annotation techniques into production processes offers opportunities for more precise control over product quality. Research, for instance [5], underscores the significance of semantic video analysis in comprehending the context of operator activities, which in turn enables the adaptation of production procedures to real-world conditions. Additionally, accounting for the particularities of human-machine interaction is essential, as emphasized in [6], which explores more intricate models for analyzing actions in industrial settings.

The current study investigates the issue of segmentation and categorization of production tools in a computerized setting using the One-Shot Learning (OSL) technique (a method in machine learning that trains a model to recognize objects or categories based on only one example per category). The primary objective is to create a system that can effectively identify and categorize tools using a minimal amount of training data.

To solve this problem, the study integrates pre-trained models for extracting feature vectors and employs semantic segmentation to enhance the accuracy of object detection in images. The use of neural network architectures, such as Siamese networks, and spatial analysis techniques play a crucial role in the successful operation of the system.

The scientific significance of this research lies in the development of an integrated architecture that combines feature extraction utilizing neural network algorithms with subsequent segmentation. Such a solution would enhance the efficiency of recognizing and incorporating new objects into a production setting, where re-training neural network models upon the addition of new data regarding objects is not feasible.

2 Combined Approaches to Identifying Keypoints and Tools in Dynamic Production Processes

In the contemporary era of industrial automation, where humans interact closely with robotic systems, it is crucial to scrutinize operator actions and recognize the tools they employ. This endeavor is essential for enhancing both the safety and efficacy of industrial processes.

Previously, in a study aimed at optimizing collaboration with collaborative robots [7], we developed a module capable of recognizing key points using ensembles of deep neural networks. This module employs a multimodal camera system to detect key points on a human operator. The system monitors the operator's movements within the work area, effectively preventing potential collisions with robots. Furthermore, this approach enables precise planning of robot trajectories and real-time synchronization of their interactions with the operator, which is crucial for successful collaboration in industrial settings.

To analyze the operator's actions, it is important to determine which tool he is currently interacting with. One of the promising methods of tool recognition in an automated production environment is the use of the OSL method in combination with semantic segmentation. This is especially true when there are insufficient resources to collect and annotate large datasets for model training. Semantic segmentation allows you to accurately identify the boundaries of objects and classify them based on the context of the work area, which allows you to effectively recognize tools among many other objects in the field of view of the robot.

Figure 1 presents the proposed architecture for a system that visualizes the operator's actions in automated industrial production settings, not only capturing the operator's movements and the tools they interact with but also creating an ontology of these actions.

The system integrates several video data streams, including RGB and depth frames, from various cameras located in the operator's work area. These inputs are essential for providing a multimodal analysis of the production environment.

The central component of the system is a neural network-based detector block, which includes key point and object detectors. Key point detectors identify significant features on the operator's body and objects with which they interact, allowing for real-time tracking of the operator's movements and actions. Additionally, an object detection unit operates in parallel, focusing on identifying and segmenting tools in the work area.

It is worth noting that the OSL method is employed in this process. This method allows the system to accurately recognize and segment tools, even with a limited number of training samples.

The data collected using detectors undergoes a process of aggregation, which enables the synchronization and integration of information from various video sources. This is crucial for creating a comprehensive understanding of the production process and accurately analyzing the operator's interactions with tools and equipment.

Based on aggregated data, the system identifies operator actions, detecting key technological operations such as painting components, and analyzing them in the context of the overall manufacturing process. Additionally, the architecture utilizes the Large Language Model Meta AI (LLaMA), which provides semantic interpretation of data, allowing for more precise conclusions about the progress of the manufacturing process.

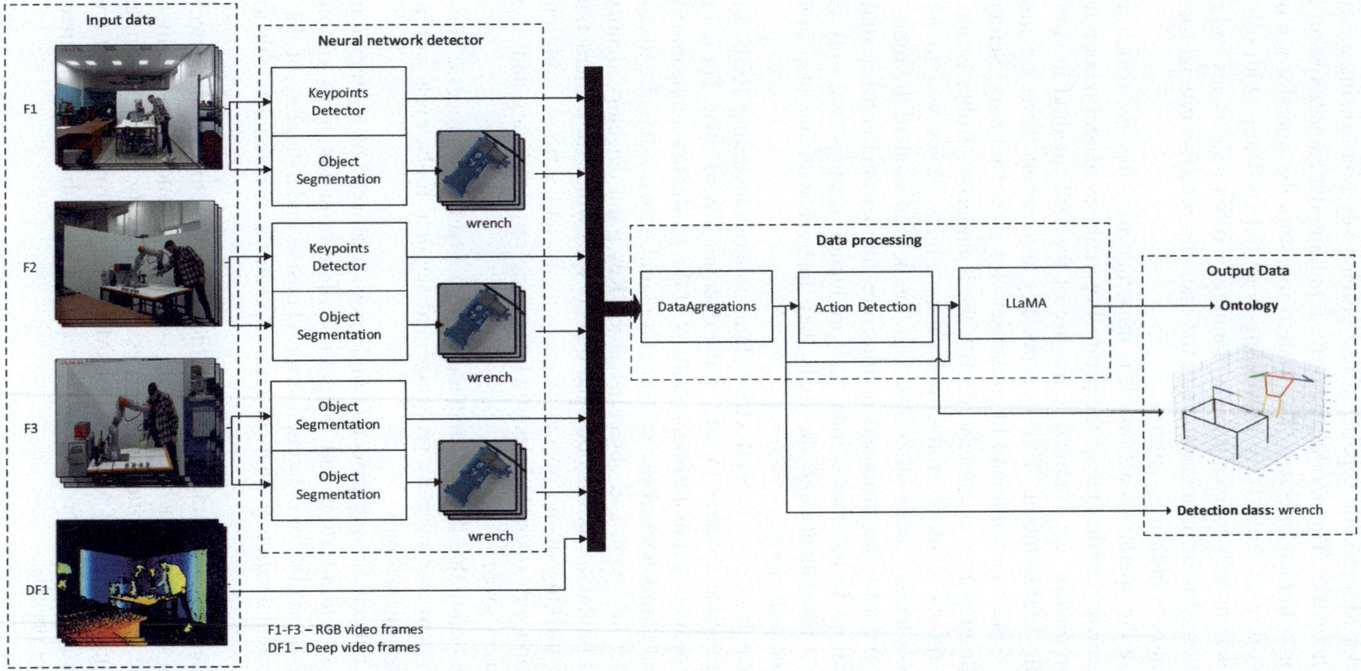

Fig. 1 Architecture of the system for detecting industrial tools and analyzing operator actions in a robotic process

As a result, the system produces output in the form of an ontology and three-dimensional movement trajectories, which facilitates the visualization and analysis of the operator's performance and the efficiency of tool usage.

3 A Mathematical Description of One-Shot Learning Using Semantic Segmentation for the Detection of Industrial Tools

The application of OSL to the task of semantic image segmentation involves training a model to classify each individual pixel in an image based on a reference example for each object class. This is accomplished by extracting feature vectors from the reference image and comparing these with pixel features in the target image.

For effective implementation, techniques are employed that combine distance metrics in high-dimensional feature spaces and specialized neural network architectures, such as siamese networks and models with an attention mechanism. These approaches allow the model to accurately segment objects, even with limited training data, by utilizing the power of deep learning algorithms.

Let $\boldsymbol{X} = \{x_1, x_2, \ldots, x_n\}$ be an image, where x_i is the feature vector corresponding to each pixel. The task of semantic segmentation is to define a class $c_i \in \zeta$ for each pixel x_i, where ζ is a set of classes. In OSL, only one example is available for each class: $\left\{x^c_{ref}, y^c_{ref}\right\}$, where x^c_{ref} is the image of an example class, y^c_{ref} is the segmentation label for the example.

The main architecture used for OSL is Siamese networks, presented in Bromley's work [8] and supplemented in [9]. In such networks, two identical paths are created for the image of the input example y^c_{ref} and the target image $\boldsymbol{X}$, through which both images pass, which makes it possible to calculate their representations in the feature space.

Let's denote the neural network function f_θ, which maps the image x to the feature space: $f_\theta : x \to \mathbb{R}^d$, where θ are the network parameters. Then the representation of each pixel x_i and pixel of the reference image x^c_{ref} in the feature space can be written as:

$$f_\theta(x_i) = z_i,\ f_\theta\left(x^c_{ref}\right) = z^c_{ref} \tag{1}$$

A key component of OSL is the choice of a metric for comparing feature vectors z_i and z^c_{ref}. One of the common options is the cosine distance:

$$d\left(z_i, z^c_{ref}\right) = 1 - \frac{z_i \cdot z^c_{ref}}{||z_i|| \cdot \left|\left|z^c_{ref}\right|\right|} \tag{2}$$

Another option may be the Euclidean distance:

$$d\left(z_i, z_{ref}^c\right) = \sqrt{\sum_{i=1}^{n}\left(z_i - z_{ref}^c\right)^2} \tag{3}$$

To train a neural network based on a single example per class, specialized loss functions can be employed. One such method is the triplet loss [10], which is a popular approach. In this method, three elements are selected: x_i is the target pixel, x_{ref}^+ is a positive example (a pixel of the same class), and x_{ref}^- is a negative example (a pixel of another class). The loss function is formulated as:

$$\mathcal{L}_{triplet} = \max\left(0, d\left(z_i, z_{ref}^-\right) - d\left(z_i, z_{ref}^+\right) + \alpha\right) \tag{4}$$

where α is the boundary that controls the minimum difference between positive and negative distances.

To improve the accuracy of segmentation, we use the attention mechanism proposed in [11]. For each pixel x_i, attention $A(i,j)$ is calculated, which determines the degree of correlation between x_i and each pixel of the reference image x_{ref}^c. The mechanism of attention can be described as:

$$A(i,j) = \frac{\exp\left(\frac{z_c \cdot z_{ref_j}^c}{\sqrt{d}}\right)}{\sum_{j'} \exp\left(\frac{z_c \cdot z_{ref_{j'}}^c}{\sqrt{d}}\right)} \tag{5}$$

where $z_{ref_j}^c$ is the feature vector of the pixel j of the reference image. The final probability that pixel x_i belongs to class c is given by the sum of all pixels of the reference image:

$$P(c|x_i) = \sum_j A(i,j) \tag{6}$$

Network training involves minimizing a loss function, which is a measure of the difference between the pixel representations of target images and reference images. Additionally, various regularization techniques can be employed, such as dropout [12] or batch normalization [13].

4 The Development of Classification Models for the Extraction of Features from Images

For the Siamese network to function, it is essential to effectively assign a feature vector. Based on this feature vector, the OSL algorithm determines the degree of similarity between presented objects. The feature vector represents a compact representation of the original image in multidimensional space. Similar objects should be positioned close to each other in this space, while dissimilar objects should be situated at a distance. This stage is crucial for ensuring the successful operation of OSL techniques that rely on differences between pairs of images during training.

In this study, several pre-trained models were fine-tuned to extract highly informative feature representations from images. The following models were selected for this purpose:

- EfficientNet B3 [14]
- ResNet50 [15]
- ConvNeXt [16]
- EfficientNet V2 [17]

The training was conducted using a custom dataset that consisted of four categories: bayonet, spray gun, welder, and wrench. For each category, 200 images were used for training and an additional 50 were used for validation (Fig. 2). The images were evenly distributed among each category to ensure a balanced sample.

Each of these models has been customized to perform the function of feature extraction utilizing the transfer learning technique. These models were trained on a fresh dataset, which enabled them to adapt to the particular features of the classes and enhance their capacity to identify significant characteristics. The architectures varied in terms of the size of input images and depth of the network, which likewise impacted their capacity for generalization and accuracy in the task of class identification (Table 1).

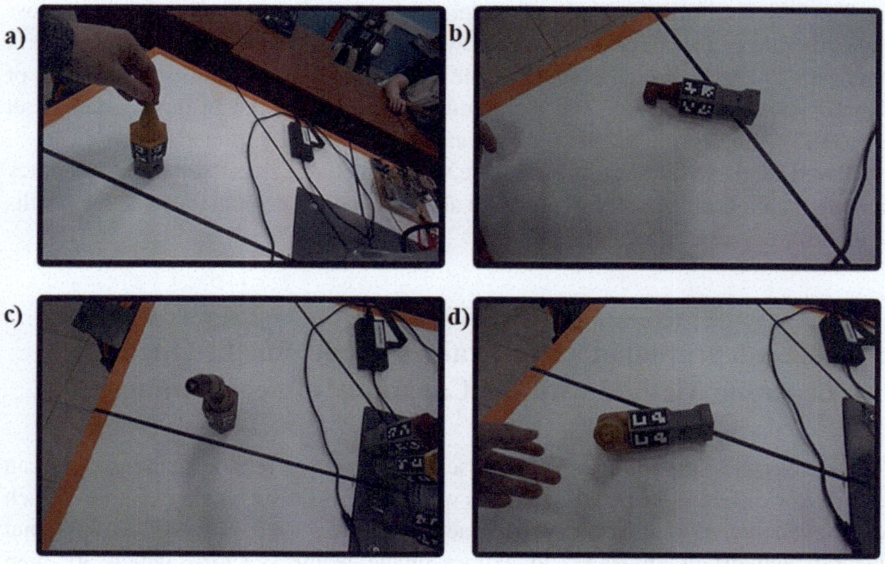

Fig. 2 Own data set for transfer training: (**a**) bayonet; (**b**) spray_gun; (**c**) welder; (**d**) wrench

Table 1 Result of training

Model	Input image size	Accuracy (train)	Accuracy (validation)
EfficientNet B3	224 × 224	0.8162	0.9141
ResNet50	224 × 224	0.9402	0.9673
ConvNeXt	200 × 200	0.9614	0.9687
EfficientNet V2	400 × 400	0.9029	0.9590

The table presents the main learning outcomes, including the input image size for each architecture and the accuracy on the validation set. As can be observed from the data, ConvNeXt demonstrated the best results, with an accuracy of 0.9687. This indicates its strong ability to generalize and identify significant features.

During each training stage, the loss functions for both the training and testing samples, as well as the accuracy metric were monitored. The figures (Fig. 3) present the results for each model in graphical form.

For the EfficientNetB3 model, the loss function during training and testing shows a rapid decrease and stabilization. Although there are slight fluctuations in the test set, they may indicate signs of overfitting. The maximum accuracy achieved in the test set was 0.9141, indicating the high performance of the model for tool classification.

Despite the slower decrease of the ResNet50 model's loss function in the test set compared to the training set, the model achieved an accuracy of 0.9673. This indicates a good generalization ability.

The ConvNeXt architecture achieved the best results among all models, with a smooth and stable reduction in the loss function. The validation accuracy was 0.9687, demonstrating the model's ability to generalize and classify tools accurately.

The EfficientNetB0V2Medium model demonstrated a rapid decrease in the loss function during the training process, however, some fluctuations were observed in the testing sample, which could indicate the presence of variability in the data or insufficient regularization. The maximum accuracy achieved in the validation set was 0.9590, indicating the good classification ability of the model.

Based on these findings, the ConvNeXt architecture exhibited the best accuracy among the models considered, although all architectures demonstrated stable results with accuracy exceeding 90%.

5 The Integration of a Pre-Trained Model for the One-Shot Learning Method with the Capability of Segmentation

To address the OSL problem, it is essential to extract reliable features that can effectively differentiate between objects when there is only a single example of each class available. In our proposed approach, we utilize a pre-trained Convolutional Neural Network (ConvNeXt) to extract image feature vectors, which are then utilized for object comparison in the OSL context.

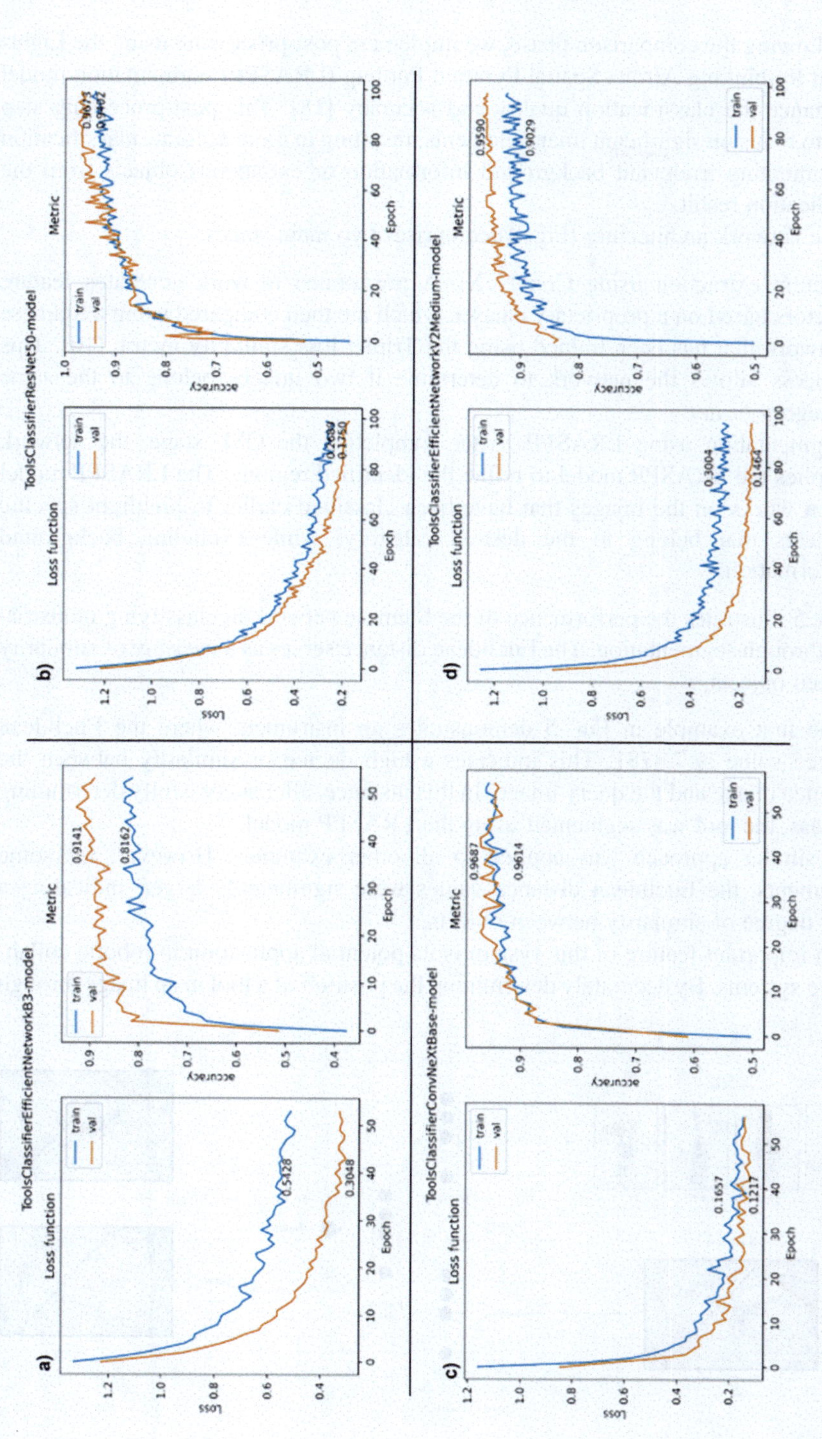

Fig. 3 Model training schedules: **(a)** EfficientNet B3; **(b)** ResNet50 **(c)** ConvNeXt **(d)** EfficientNet V2

Following the comparison phase, we implement post-processing using the Lightweight Rethinking Atrous Spatial Pyramid Pooling (LRASPP) segmentation model to enhance the classification quality and accuracy [18]. This post-processing step helps to focus on significant image elements, resulting in more accurate classification by eliminating irrelevant background information or extraneous objects from the classification result.

The network architecture (Fig. 4) comprises two main stages:

1. Feature extraction using ConvNeXt. A pre-trained network generates feature vectors based on a proprietary dataset, which are then compared using a Siamese network that has been trained using the Triplet loss similarity metric [10]. This process allows the network to determine if two images belong to the same category or not.
2. Segmentation using LRASPP. After completing the OSL stage, the network applies the LRASPP model to refine the identified regions. The LRASPP model then works on the images that have been classified earlier to highlight specific objects that belong to the desired category, while excluding background information.

Figure 5 illustrates the performance of the Siamese network in classifying industrial tools through segmentation. The Euclidean distance serves as a measure of similarity between objects.

The first example in Fig. 5 demonstrates an instrument where the Euclidean distance value is 7.8781. This indicates a high degree of similarity between the reference image and the query image. In this instance, after successfully determining the class, the tool was segmented using the LRASPP model.

A similar approach was applied to all other examples. However, for some instruments, the Euclidean distance values were significantly larger, indicating a lesser degree of similarity between objects.

An important feature of this system is its potential application in robotic collaborative systems. By accurately determining the position of a tool in an image through

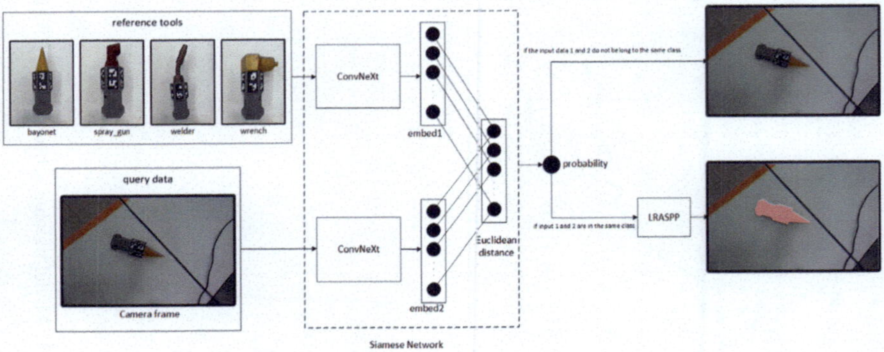

Fig. 4 One-shot learning architecture with semantic segmentation of objects

Fig. 5 Experimental results of the OSL method in conjunction with the LRASPP model

segmentation, robots can interact more effectively with the tool and perform production tasks with greater precision.

6 Conclusion

In this research, a combined architecture has been developed for segmentation and classification of industrial tools in an automated setting using the OSL method. The primary objective of this work was to develop a system that can effectively identify and segment tools based on a minimal amount of training data. This is particularly relevant in scenarios where retraining neural network-based algorithms when introducing new objects is not feasible.

During this study, several models (EfficientNet B3, ResNet50, ConvNeXt, and EfficientNet V2) were trained on their own custom dataset that includes four categories of tools. This allowed us to obtain feature vectors that are essential for the OSL technique to function. The ConvNeXt architecture demonstrated the best performance with an accuracy of 96.87% on the validation set, indicating its strong ability to generalize and extract significant features.

Despite the results achieved, the system still has the potential for improvement. In particular, the segmentation component could be enhanced by employing more advanced techniques. A promising avenue for improvement is the application of the approach proposed in [19], which provides a general method for segmenting any object without the need for dedicated training on specific classes.

Acknowledgments The work was supported by the Russian Science Foundation (project No. 22-71-10093) [20].

References

1. Nevatia, R., Hobbs, J., Bolles, B.: An ontology for video event representation. In: Conference on Computer Vision and Pattern Recognition Workshop, pp. 119–119. IEEE (2004)
2. Kanimozhi, S., Sasithradevi, A., Sairamesh, L.: Ontology based action recognition in sport video's using semantic verification model. IEEE Access. **12**, 96783–96796 (2024)
3. Yan, P.: Prompt engineering-assisted malware dynamic analysis using gpt-4. arXiv preprint arXiv:2312.08317 (2023)
4. Ballan, L.: Event detection and recognition for semantic annotation of video. Multimed. Tools Appl. **51**, 279–302 (2011)
5. Bertini, M., Del Bimbo, A., Serra, G.: Learning ontology rules for semantic video annotation. In: Proceedings of the 2nd ACM Workshop on Multimedia Semantics, pp. 1–8. ACM (2008)
6. Youssef, W.F., Haidar, S., Joly, P.: Generic video surveillance description ontology. In: BDCSIntell, pp. 83–88 (2018)
7. Ivanov, Y.: Using an ensemble of deep neural networks to detect human keypoints in the workspace of a collaborative robotic system. Eng. Proc. **33**, 19–24 (2023)

8. Bromley, J.: Signature verification using a "siamese" time delay neural network. In: Advances in Neural Information Processing Systems, vol. 6 (1993)
9. Koch, G.: Siamese neural networks for one-shot image recognition. In: ICML Deep Learning Workshop, vol. 2, pp. 1–30 (2015)
10. Schroff, F., Kalenichenko, D., Philbin, J.: Facenet: A unified embedding for face recognition and clustering. In: Proceedings of the IEEE Conference on Computer Vision and Pattern Recognition, pp. 815–823. IEEE (2015)
11. Vaswani, A.: Attention is all you need. In: Advances in Neural Information Processing Systems (2017)
12. Srivastava, N.: Dropout: a simple way to prevent neural networks from overfitting. J. Mach. Learn. Res. **15**, 1929–1958 (2014)
13. Ioffe, S.: Batch normalization: Accelerating deep network training by reducing internal covariate shift. arXiv preprint arXiv:1502.03167 (2015)
14. Tan, M.: Efficientnet: Rethinking model scaling for convolutional neural networks. arXiv preprint arXiv:1905.11946 (2019)
15. He, K.: Deep residual learning for image recognition. In: Proceedings of the IEEE Conference on Computer Vision and Pattern Recognition, pp. 770–778. IEEE (2016)
16. Liu, Z.: A convnet for the 2020s. In: Proceedings of the IEEE/CVF Conference on Computer Vision and Pattern Recognition, pp. 11976–11986. IEEE (2022)
17. Tan, M., Le, Q.: Efficientnetv2: smaller models and faster training. In: International Conference on Machine Learning, pp. 10096–10106. PMLR (2021)
18. Howard, A.: Searching for mobilenetv3. In: Proceedings of the IEEE/CVF International Conference on Computer Vision, pp. 1314–1324. IEEE (2019)
19. Kirillov, A.: Segment anything. In: Proceedings of the IEEE/CVF International Conference on Computer Vision, pp. 4015–4026. IEEE (2023)
20. RSCF Homepage. https://rscf.ru/project/22-71-10093/. Accessed 22 Aug 2024

The Kolmogorov-Smirnov Mathematical Test for Vectors of the Digital Twin of an Enterprise

Sergey Masaev, Ivan Chumakov, and Valentina Vingert

Abstract The research analyzes the use of the Kolmogorov-Smirnov mathematical test to identify the statistical significance of the vectors of the digital twin of the enterprise. The theoretical foundations of this test, including its mathematical formulation and application conditions, have been studied. Criteria for the significance of vectors are defined, taking into account the specifics of the data and the goals of the analysis. The null and alternative hypotheses regarding the significance of the selected set of vectors are formulated. The maximum vertical deviation between the empirical data distribution function and the theoretical distribution is calculated. The critical estimate of the level of significance of the test for making decisions about hypotheses has been established. The analysis of the results with conclusions about the influence of vectors on the modeling and optimization of the system is carried out.

Keywords Kolmogorov-Smirnov test · Normal distribution · Hypothesis · Critical value · Maximum deviation · Digital twin

1 Introduction

Modern enterprises are increasingly aware of the importance of using digital twins to optimize their operations. A digital twin is a virtual model of a real object or system that represents its state, behavior and properties in real time. Enterprises are able to

S. Masaev
Reshetnev Siberian State University of Science and Technology, Krasnoyarsk, Russia

Siberian Federal University, Krasnoyarsk, Russia

I. Chumakov
Reshetnev Siberian State University of Science and Technology, Krasnoyarsk, Russia

V. Vingert (✉)
Siberian Federal University, Krasnoyarsk, Russia

A. Gibadullin, S. Gordin (eds.), *Computing Technologies and Applied Mathematics*, Springer Proceedings in Mathematics & Statistics 500,
https://doi.org/10.1007/978-3-031-90546-9_14

conduct various types of analysis, simulations and optimizations without interfering with real processes by creating and using digital twins.

It is important to ensure its accuracy and reliability, then the digital twin will bring maximum benefit. Digital twin vectors play a crucial role in the creation and analysis of these models. Each vector is a set of numerical values that describes the state and characteristics of a real object or system at a specific point in time. Vectors play an important role in modeling and analyzing digital twins, since they allow complex data to be presented in a form that is easy to process.

Vector creation begins with collecting data on various parameters of an object or system. This data may include physical measurements, performance indicators, temperature conditions, and other relevant variables. All these parameters are then converted into numerical form and combined into a single vector that can be used for further analysis [1].

The Kolmogorov-Smirnov test can be used to check the statistical consistency and detect anomalies of the digital twin vectors. This statistical test provides an opportunity to estimate the degree of compliance of the sample with a given theoretical distribution. Applying the Kolmogorov-Smirnov test to a set of vectors can reveal inconsistencies and anomalies that may indicate problems or errors in the enterprise digital twin.

The creation of the digital twin of the enterprise and its application in various fields is widely covered in scientific works, including those by Myasnikova O.V., Prokhorov A.A., Zhilyaev A.A., Vasilyev A.N. and others. The description of the processes occurring on the digital twin with the help of vectors can be found in the field of computer science, informatics and artificial intelligence. Socher R., Pennington J., Mikolov T. and others. There is a substantial body of research articles, books and journal publications devoted to this topic. These sources typically describe methods and algorithms that use vector representation to analyze and process data on a digital twin. Despite the extensive research literature published on the subject, this research focuses on the under-examined problem of using the Kolmogorov-Smirnov test to test the significance of a set of vectors of an enterprise digital twin. In this paper, we examine the application of the Kolmogorov-Smirnov test to a set of vectors of the digital twin of an enterprise. These scientific and technical sources often provide detailed descriptions of methods and algorithms that use vector representation of data to conduct comprehensive analysis and processing of information on the digital twin platform. These methods allow for efficient modeling of complex systems and processes, providing a deeper understanding of their functioning and revealing hidden patterns. Algorithms using vector representation offer tools for optimizing calculations and increasing the accuracy of simulations, which is especially important when working with large volumes of data. Despite the abundance of scientific literature on this topic, this research focuses on the understudied problem of using the Kolmogorov-Smirnov test to estimate the significance of a set of vectors of an enterprise digital twin [2].

This paper will consider the use of the Kolmogorov-Smirnov test to analyze a set of vectors in order to identify their statistical significance.

The purpose of this research is to establish the statistical significance of the vectors of the digital twin of the enterprise using the Kolmogorov-Smirnov test.

To achieve this research, we completed the following tasks:

Explore the theoretical basis of the Kolmogorov-Smirnov test and its application to hypothesis testing. It is necessary to examine the theoretical aspects underlying the Kolmogorov-Smirnov test, including its mathematical formulation and application conditions. It is important to understand how this test is used to estimate the agreement between the empirical distribution of data and the assumed theoretical distribution when testing scientific hypotheses.

Define criteria for the significance of enterprise digital twin vectors. Clear criteria and indicators should be developed to estimate the significance of specific vectors representing the states and parameters of the enterprise digital twin. These criteria should take into account the specifics of the data under consideration and the objectives of the analysis.

Formulate the null and alternative hypotheses regarding the significance of the selected set of vectors. It is necessary to formulate the null hypothesis (usually stating the absence of statistically significant differences) and the alternative hypothesis (assuming the presence of such differences) to analyze the significance of the vectors under consideration.

Prepare data for analysis by identifying the necessary vectors. At this stage, it is important to carefully select and prepare the data that will be used for the analysis, paying special attention to identifying those vectors that are central to the re-search.

Calculate the maximum vertical deviation between the empirical distribution function of the data and the theoretical distribution function. It is necessary to calculate the maximum difference between the cumulative empirical frequency of the data and the theoretically expected frequency at all possible levels of the distribution, which is the key indicator in the Kolmogorov-Smirnov test.

Determine the critical value. It is important to determine the critical value for the significance level of the selected test, which will serve as the threshold for making a decision to support or reject the null hypothesis.

Analyze the results and draw conclusions about the significance of the selected set of vectors. After all calculations have been completed, it is necessary to interpret the results obtained and draw reasonable conclusions regarding the statistical significance of the analyzed vectors of the enterprise digital twin, determining their impact on the overall process of modeling and optimizing the system.

2 Method

The Kolmogorov-Smirnov test is one of the most commonly employed statistical tests for evaluating the hypothesis that an empirical distribution conforms to a theoretical distribution. This test is based on the comparison of the empirical distribution function (EDF) and the distribution function (DF) of the theoretical distribution. It allows one to determine how accurately the empirical distribution matches the theoretical distribution.

Prior to an examination of the Kolmogorov-Smirnov test itself, it is essential to re-view some fundamental concepts and definitions that will be employed subsequently.

The Kolmogorov-Smirnov test is founded upon a comparison between two distributions: empirical and theoretical. The empirical distribution represents the actual data obtained through observation or experiment. In contrast, the theoretical distribution is a mathematical model that describes the anticipated behavior of a random variable.

The distribution function is the principal instrument for delineating the distribution of a random variable. It enables the probability of a random variable assuming a value of less than or equal to a specified value to be determined. The distribution function is typically represented by the symbol F(x) and takes the following form:

$$F(x) = P(X \leq x) \tag{1}$$

where X is the random variable under study.

The empirical (sample) distribution function (EDF) is a statistical method used to determine the probability distribution function of a random variable based on available data. It is written as follows:

$$F_n(x) = \frac{n_x}{n} \tag{2}$$

n_x—the number of observations (options) is less x;

n—sample size;

EFR is a non-decreasing function, a non-negative function between zero and one:

$$0 \leq F_n(x) \leq 1 \tag{3}$$

at

$$F_n(-\infty) = 0, F_n(+\infty) = 1 \tag{4}$$

A hypothesis is a statement about the distribution parameters of a random variable or about the relationship between several random variables [3].

The significance level is the critical value of probability, above which the null hypothesis can be rejected in a statistical test. It determines the risk of making a type I error, that is, a false rejection of the true null hypothesis. The value usually used is 0.05 or 5%, which means accepting a 5% probability of an erroneous conclusion about the existence of an effect or difference, denoted by α.

P-value (critical value) is the probability that the observed test results will occur if the null hypothesis is true. P-value is used to decide whether to accept or reject the null hypothesis. If the P-value is less than the significance level, then the null hypothesis is rejected.

Now let's look at the use of the Kolmogorov-Smirnov test in more detail.

This test is based on the comparison of the empirical cumulative distribution function (ECDF) with the theoretical cumulative distribution function (TCDF). The ECDF is the accumulated fraction of values less than or equal to a given value. The TFR, in turn, is an analogue of the distribution function for the theoretical distribution [4].

The basic idea of the Kolmogorov-Smirnov test is to compare the maximum deviation between the EFR and the TFR. If this deviation is small, we can conclude that the empirical distribution corresponds to the theoretical distribution. Otherwise, the hypothesis that the distributions match is rejected.

In this article, we will look at the use of the Kolmogorov-Smirnov test to analyze data presented as digital twin vectors. This analysis will also help to identify key criteria that are necessary for optimizing digital twin models. Such important criteria include the relevance of data to current tasks, the accuracy of information presentation, the impact on the overall functioning of the real enterprise, compliance with the company's strategic goals, and the amount of data required to conduct a quality analysis.

To estimate the significance of a selected data set in the context of an organization's digital twin using the Kolmogorov-Smirnov test, several sequential steps must be completed:

Collecting and organizing data: First, it is necessary to collect all the data that will be used to test whether it fits the given theoretical distribution.

Statement of hypotheses: The null and alternative hypotheses should be clearly stated. The null hypothesis assumes that the observed empirical distribution does not differ from the theoretical distribution, while the alternative hypothesis states the opposite.

Calculation of the empirical distribution function (EDF): The EDF is the cumulative proportion of observations equal to or less than each value in the data set.

Selecting a theoretical distribution: It is necessary to select a suitable theoretical distribution that will serve as a basis for comparison with the available data. Then, its theoretical distribution function (TDF) must be established.

Calculating deviations: For each value in the sample, the difference between the EGF and the TGF should be calculated.

Determination of the maximum deviation: An important step is to calculate the maximum difference between the empirical distribution function of the EDF (Fn(x)) and the theoretical distribution function of the TDF (F(x)). This value is designated as Dn and is formally determined by the formula:

$$D_n = sup|F_n(x) - F(x)| \tag{5}$$

Critical value: in the Kolmogorov-Smirnov test it depends on the significance level (α) and the sample size (n). The critical value can be determined either from statistical tables of critical values or approximated by special formulas, such as:

$$D_{\alpha} = \frac{\sqrt{-0.5 \cdot ln\left(\frac{\alpha}{2}\right)}}{\sqrt{n}} \tag{6}$$

For more accurate calculations, we will use statistical tables that take into account the level of significance and the volume of data.

Comparison of values: The obtained value of the Kolmogorov-Smirnov statistic is compared with the critical value from the table or software.

Decision Making: Based on the comparison of the Kolmogorov-Smirnov statistic value and the critical value, a decision is made to accept or reject the null hypothesis. If the statistic value is less than the critical value, the null hypothesis is rejected, indicating that the empirical distribution does not match the theoretical distribution.

Thus, following these steps will allow you to thoroughly analyze the digital twin data and make informed conclusions regarding its significance and impact on the enterprise's model processes.

To successfully conduct the Kolmogorov test, it is recommended to use samples of at least 50 observations. This helps to reduce the impact of random variations on the results of the analysis and, therefore, increases the reliability of the test. At the same time, the number of criteria should remain small (optimally up to 10). This approach allows you to identify the category in which the sum of the frequencies of discrepancies between the two analyzed distributions reaches maximum values. In addition, it helps to estimate the statistical significance of this discrepancy in order to determine the degree of normality of the distribution of the studied populations. As a result, such an analysis contributes to a better understanding of the characteristics of the data and their role in the general enterprise management system [5].

Two types of data were employed to estimate the significance of the investigated dataset in the context of the digital twin of the enterprise:

The first type is data obtained using the enterprise digital twin management method through the goal tree. This goal-setting method allows you to clearly formulate and organize the company's goals, from its mission and strategic plans to specific tasks. Using the goal tree not only makes it possible to see the relationships between different goal levels, but also helps to analyze the impact of changes at all management levels [6].

The second type of data is data collected without using this method.

The vectors represent the processes of the digital twin of the enterprise (X1, X2, X3, X4). X4 is the sum of X1, X2 and X3. In this instance:

X1 represents a non-stationary process. A non-stationary process is defined as a process exhibiting statistical characteristics that undergo temporal variation. Such changes may include alterations in the mean, variance, correlation, and so forth.

X2 is conditionally stationary. A process is said to be conditionally stationary (or stationary in the broad sense) if it is not strictly stationary, but its first and second moments (mean and autocorrelation function) do not change with time. This implies that although the process may undergo fluctuations, its fundamental statistical characteristics remain unchanging.

X3 is stationary. A stationary process is defined as a random process whose statistical characteristics remain constant over time. For such processes, the mean value and the autocorrelation function remain constant.

Thus, using these data and taking into account the differences in the characteristics of the processes (stationarity or lack thereof), it is possible to more deeply understand the behavior of the digital twin of the enterprise and draw conclusions about the influence of the studied factors on the general management system of the company.

Let us formulate hypotheses that will be tested using the Kolmogorov-Smirnov test:

First hypothesis (H0): This hypothesis assumes that the vector data correspond to a normal distribution. The correspondence is confirmed if the calculated value of the maximum deviation between the empirical distribution function (EDF) and the theoretical distribution function (TDF) does not exceed a predetermined critical value. This means that the statistical characteristics of the sample do not contradict the normality of the data distribution.

Second hypothesis (HA): The alternative hypothesis states that the vector data do not obey the normal distribution. This will be proven if the maximum deviation value is greater than the established critical value. Such a discrepancy indicates that the observed data differ significantly from the values that could be obtained under the condition of normal distribution.

To test these hypotheses, it is necessary to consistently calculate the values of the empirical distribution function and the theoretical distribution function for each of the eight vectors under consideration, representing different processes. After that, it is necessary to calculate the maximum deviation between the EDF and the TDF for each vector and determine the Kolmogorov-Smirnov statistics. All obtained values of the maximum deviations are entered into Table 1 for a visual presentation of the analysis results and further comparison with the critical values. This process will allow an objective estimation of the conformity of the data to the normal.

The Kolmogorov-Smirnov statistic is then compared with the critical value obtained from the table. For $n = 60$ and $\alpha = 0.05$, the critical value is 0.17533 [7].

For illustrative purposes, we have plotted the empirical and theoretical distributions of processes X1, X2, X3, X4, with and without the application of the method (see Figs. 1 and 2).

Since the maximum deviation values are significantly greater than the critical value, the null hypothesis must be rejected. This means that our sample data does not fool-low a normal distribution. This result indicates that there are significant differences between the empirical and theoretical distributions, indicating that the data is not normal [8].

Table 1 Maximum deviations

Xn	X1	X2	X3	X4
With method	0.6466	0.7701	0.9667	0.9666
No method	0.6394	0.771	0.9667	0.9666

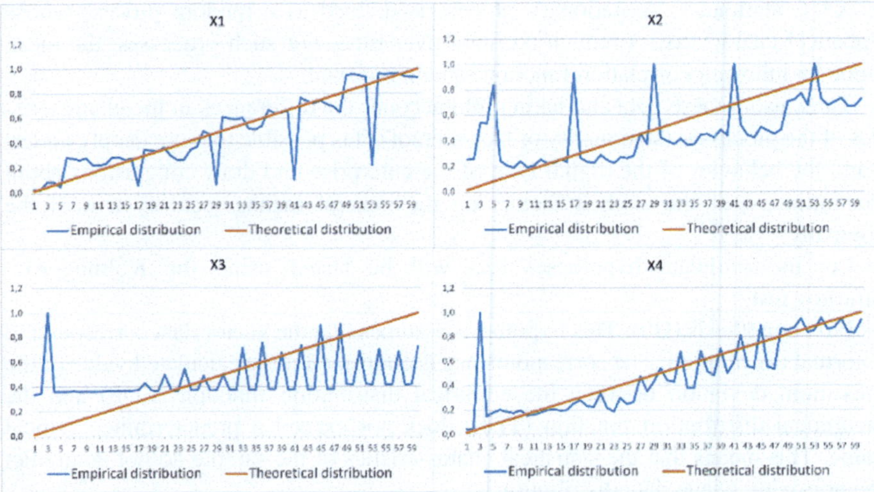

Fig. 1 Graphs X1, X2, X3, X4 using the method

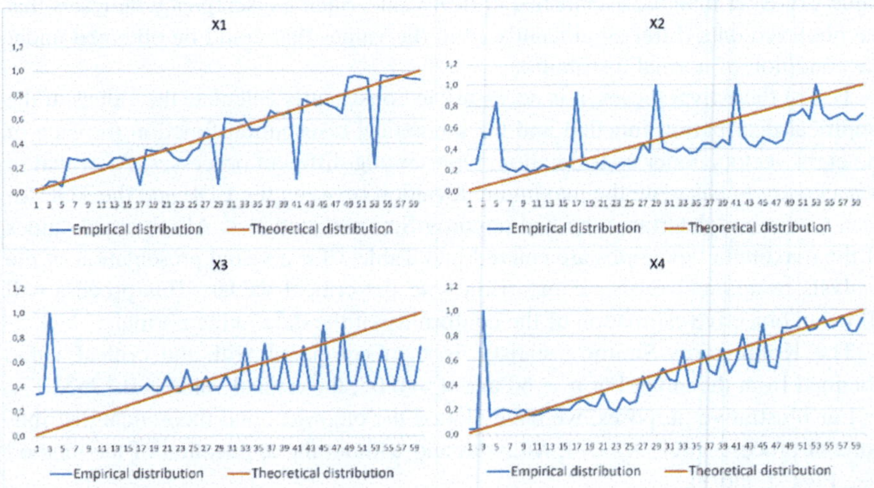

Fig. 2 Graphs X1, X2, X3, X4 without using the method

It is also worth noting that the graphs of processes X1, X2, X3 and X4 demonstrate a certain periodicity in the time scale with pronounced vertical spikes up to maximum values. These observations suggest the presence of cyclicity in the processes, which may be caused by external or internal factors influencing their dynamics.

We decide to choose another verification criterion for a more detailed analysis. Within this approach, we divide the available data into groups depending on their

Table 2 Maximum deviations for X1

With method	0.4107	0.5522	0.3995	0.5503	0.593
No method	0.4099	0.5448	0.3976	0.5592	0.5976

Table 3 Maximum deviations for X2

With method	0.6786	0.6659	0.6659	0.6659	0.6659
No method	0.6788	0.6659	0.6659	0.6659	0.6659

Table 4 Maximum deviations for X3

With method	0.8326	0.5377	0.5489	0.5652	0.7276
No method	0.8326	0.5356	0.5475	0.5646	0.7244

Table 5 Maximum deviations for X4

With method	0.8324	0.6905	0.502	0.5685	0.7359
No method	0.8324	0.6965	0.4925	0.5687	0.7359

belonging to a certain year. Each such group, or segment, will include five processes for detailed consideration.

Using the selected new criterion and applying a sequence of actions to calculate statistical characteristics, we obtain the values of the maximum deviations for each group of data. These results are presented in Tables 2, 3, 4 and 5. This approach allows us to more accurately estimate the behavior of the data over specific periods of time and identify possible trends or anomalies in their distribution.

To enable a more visual and detailed comparison of the results of the analysis, graphs were constructed showing the distributions of the empirical data in comparison with the theoretical distributions. These visual representations help to better understand the nature of the differences between the observed data and the theoretical model, as well as to identify potential causes of the discrepancies.

Figure 3 shows graphs for processes X1, X2, X3 and X4, which were obtained using the goal tree management method. These graphs allow us to trace the dynamics of changes in each process over time and estimate the degree of their compliance with the theoretical distribution. Such visualization also helps to identify possible patterns or deviations in the behavior of processes.

Note that the process graphs without the use of this method do not show significant differences from those constructed using the method. This suggests that the basic nature of the data distribution is preserved regardless of the use of this approach. However, the use of the method can still provide additional advantages in terms of structuring and analyzing data, especially when processing large volumes of information.

When conducting the analysis and dividing the data into groups with the number of observations $n = 12$, the critical value adopted for testing the hypothesis of

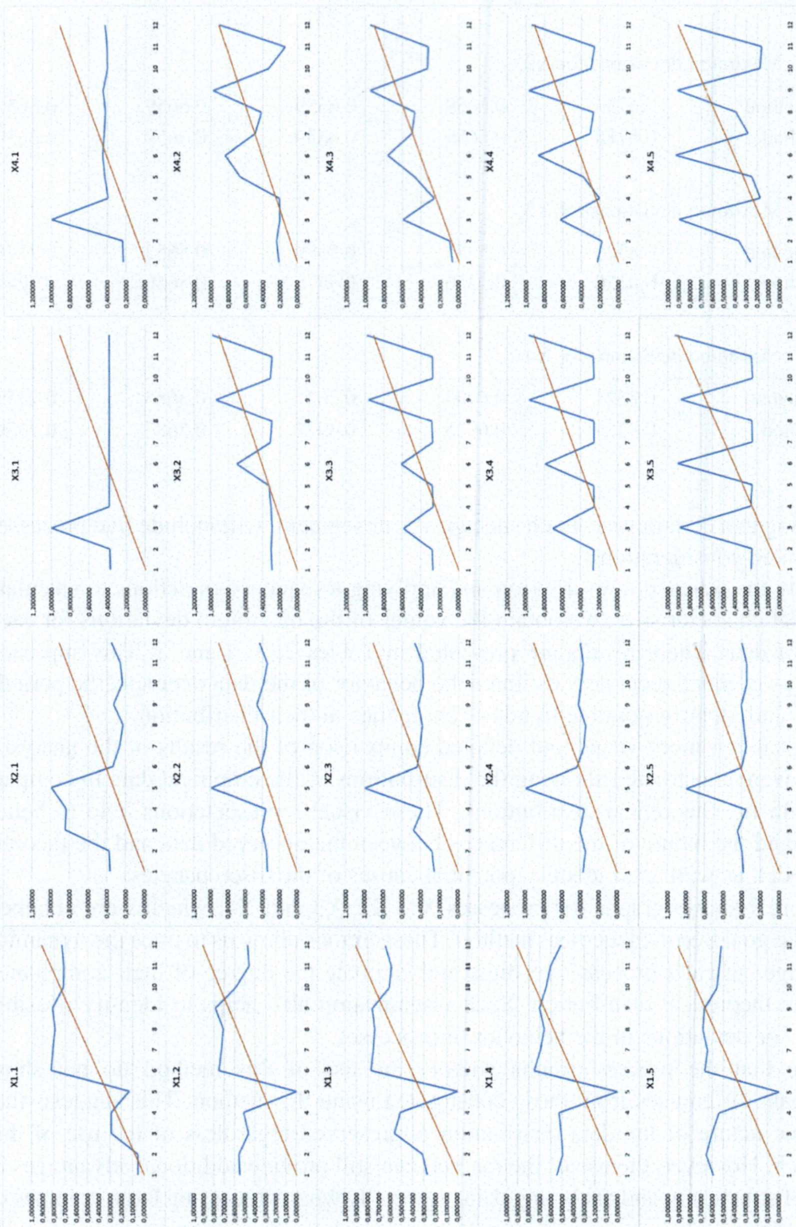

Fig. 3 Graphs of processes X1, X2, X3, X4 using the method

normal distribution is 0.392. This value is a threshold and serves to determine the compliance of the empirical data with the proposed theoretical model.

After performing all the necessary calculations, it was found that the actual values of the deviations significantly exceed this critical value. This fact indicates that the processes X1, X2, X3 and X4 in this case also do not correspond to the normal distribution. Such a discrepancy indicates significant differences between the real data and the normal model.

Thus, regardless of the approach to analysis and the method of grouping data, the maximum deviations are significantly higher than the established significance level. This means that the null hypothesis must be rejected in both the first and second cases of analysis. That is, these processes do not have the characteristics of a normal distribution [9].

During the research it became apparent that the criteria used clearly show deviation from normality. This may require further investigation into the causes of such abnormality and the development of alternative models or methods to more accurately describe the behavior of processes X1, X2, X3 and X4 within the framework of the enterprise digital twin tasks.

The Kolmogorov-Smirnov test helps analyze digital twin data by estimating its compliance with expected distributions. This identifies key parameters for the effectiveness of models and management. The test focuses on meaningful data, avoiding redundancy, and visualization of results improves understanding and involvement in decision making. Correct application of the test optimizes business processes and supports the company's strategic goals [10].

3 Results

During further analysis of the data, ordered by time, a certain pattern was revealed:

The graphs of processes X1 and X2 mutually reflect each other, which may indicate their insignificant influence on X4 (the sum of X1, X2, X3);

The process X3, in turn, significantly influences X4 (the sum of X1, X2, X3);

The changes in the values are seasonal, with increases and decreases and sharp vertical spikes to maximum values at certain times of the year, which indicates cyclicity in their distribution;

Presumably, the data in processes X3 and X4 are distributed, during the year, according to some periodic function, which must be clarified using more complex mathematical tools.

4 Discussion

The study of the vectors (processes) of the digital twin of the enterprise using the Kolmogorov-Smirnov test is of scientific value, as it allows for the accurate estimation of the conformity of the data of real processes of the enterprise to the theoretical normal distribution. This allows for the following:

Identify anomalies and non-standard processes that may indicate potential problems or opportunities for optimization.

Estimate the degree of variability in processes, which is important for planning and quality control.

Prepare more accurate mathematical models for forecasting and simulation, which can improve strategic planning and operational management.

The Kolmogorov-Smirnov test can be employed to estimate the hypothesis that the distribution of process vectors of the digital twin of the enterprise aligns with the anticipated theoretical distribution. This is of particular importance in the following scenarios:

Model validation: The Kolmogorov-Smirnov test can be employed to validate the fit of the digital twin processes to real data, which is of critical importance to the accuracy of predictions and analysis.

Process optimization: This test can be employed to ascertain whether alterations to processes align with the anticipated theoretical models, thereby facilitating the optimization and enhancement of production processes.

Quality control: The test ensures that processes remain within the established standards and that any changes to processes do not result in unwanted distortions.

This test is particularly useful when the shape of the data distribution is unknown or when the data does not follow standard distributions such as normal or exponential. It is therefore an ideal tool for dealing with real, often complex data that may arise in digital twin processes.

Checking for normality of the distribution of digital twin vectors is a necessary step to:

1. Ensure that the statistical analysis is correct.
2. Correctly interpret the results.
3. Select adequate prediction methods.

Ultimately, this enhances the reliability of the conclusions and the accuracy of the predictions derived from the digital twin data.

5 Conclusions

In this paper, the Kolmogorov-Smirnov test was applied to analyze a set of vectors of the digital twin of an enterprise.

The application of the Kolmogorov-Smirnov test to a set of vectors of the digital twin of an enterprise allows for the estimation of the similarity between two sets of data and the identification of potential outliers. This is particularly pertinent in the context of analyzing vectors of the digital twin of an enterprise, as such data may contain information about the processes of their functioning and efficiency.

The results and interpretation of the Kolmogorov-Smirnov test for a set of vectors of the digital twin of an enterprise permit the drawing of conclusions regarding the similarity or difference between the data sets. If the value of the Kolmogorov-Smirnov statistic is less than the critical value, it can be concluded that there is similarity between the datasets. In the event that the value of the Kolmogorov-Smirnov statistic exceeds the critical value, it can be concluded that there are significant deviations present.

The Kolmogorov-Smirnov test offers a number of advantages when used to analyze vectors of the digital twin of an enterprise. These advantages include its simplicity and versatility. The test can be applied to the analysis of different types of data, and does not require prior normalization or data transformation. Furthermore, the Kolmogorov-Smirnov test enables the estimation of the similarity between two data sets without the necessity to consider their dimensions.

However, it should be noted that the Kolmogorov-Smirnov test is not without limitations, which must be taken into account when applying it. Firstly, the test is non-parametric, which means that it does not take into account the distribution of the data and may yield erroneous results if the assumptions of normality of the distribution are violated. Secondly, the Kolmogorov-Smirnov test requires a substantial quantity of data to provide reliable estimation of the similarity or discrepancy between data sets.

Consequently, the Kolmogorov-Smirnov test represents an efficacious instrument for the analysis of vectors associated with the digital twin of an enterprise. The test enables the estimation of the similarity or difference between datasets and the identification of potential outliers. However, it should be noted that the Kolmogorov-Smirnov test has certain limitations, including its non-parametric nature and the requirement of a large amount of data. In conclusion, the Kolmogorov-Smirnov test can be a valuable tool for analyzing enterprise digital twin vectors and can assist in identifying issues and enhancing enterprise performance.

References

1. Myasnikova, O.V.: Intellectual diary of production-logical system: postprovision methodology and application for system optimisation. Bus. Innov. Econ. **7**, 103–116 (2023)
2. Mikolov, T., Chen, K., Corrado, G., & Dean, J.: Efficient evaluation of word representations in vector space. In: Proceedings of the ICLR Workshop (2013)
3. Gmurman, V.E.: Theory of Probability and Mathematical Statistics: Textbook. Vysshaya Shkola, Moscow (2003)
4. Kolmogorov, A.N., Gnedenko, B.V.: Limit Distributions for Sums of Independent Random Variables. Gostekhizdat, Moscow, Leningrad (1949)
5. Kozlov, M.V., Prohorov, Y.V.: Introduction to Mathematical Statistics. MGU, Moscow (1987)
6. Bulatov, Y.N., Kryukov, A.V.: Algorithms for constructing a digital twin of a distributed generation plant. J. Siberian Fed. Univ. Eng. Technol. **6** (2020)
7. Smirnov, N.: Table for estimating the goodness of fit of empirical distributions. Ann. Math. Stat. **19**(2), 279–281 (1948). https://doi.org/10.1214/aoms/1177730256

8. Marsaglia, G., Tsang, W.W., Wang, J.: Evaluating Kolmogorov's distribution. J. Stat. Softw. **8** (18), 1–4 (2003). https://doi.org/10.18637/jss.v008.i18
9. Morozov, N.A.: Statistical Hypothesis Testing: Guidelines. Orenburgskij gos. unt, Orenburg (2019)
10. Markov, N.A.: Maximum Values of Random Variables: Theory and Practice. Izdatel'stvo MGU, Moscow (2009)

The manufacturer's authorised representative in the EU is Springer Nature Customer Service Centre GmbH, Europaplatz 3, 69115 Heidelberg, Germany. If you have any concerns regarding our products, please contact ProductSafety@springernature.com

Printed and bound by CPI Group (UK) Ltd, Croydon, CR0 4YY
01/10/2026
02225864-0005